INNOVATION FOR DEVELOPMENT IN AFRICA

This book uncovers the many ways in which innovations and innovation system development policies have become crucial to development policy formation across Africa.

As new instruments, actors and tools emerge in development cooperation, the role of innovation in the societal development of developing countries needs to be addressed fully. This book delves into subjects as diverse as the changing development policies between the Global North and South, the role of innovation in international aid and development policies, the role of public, private and non-governmental sectors, universities and other development actors, and the potential for inclusive innovation in local communities. In particular, the book asks who benefits from innovation-focussed development policies, and if and how practical innovation instruments include the global poor.

Written in an accessible and engaging style, the book includes a range of discussion questions and further reading suggestions to suit a range of readers, from students right through to policy makers and practitioners, or anyone else looking for an introduction to innovation policies and development in Africa.

Jussi S. Jauhiainen is Professor of Geography, University of Turku (Finland) and Visiting Professor of Human Geography at the University of Tartu (Estonia).

Lauri Hooli is University Lecturer in Human Geography at the University of Turku (Finland).

Rethinking Development

Rethinking Development offers accessible and thought-provoking overviews of contemporary topics in international development and aid. Providing original empirical and analytical insights, the books in this series push thinking in new directions by challenging current conceptualizations and developing new ones.

This is a dynamic and inspiring series for all those engaged with today's debates surrounding development issues, whether they be students, scholars, policy makers and practitioners internationally. These interdisciplinary books provide an invaluable resource for discussion in advanced undergraduate and postgraduate courses in development studies as well as in anthropology, economics, politics, geography, media studies and sociology.

Participatory Arts in International Development
Edited by Paul Cooke and Inés Soria-Donlan

Energy and Development
Frauke Urban

Power, Empowerment and Social Change
Edited by Rosemary McGee and Jethro Pettit

Innovation for Development in Africa
Jussi S. Jauhiainen and Lauri Hooli

INNOVATION FOR DEVELOPMENT IN AFRICA

Jussi S. Jauhiainen and Lauri Hooli

Routledge
Taylor & Francis Group

LONDON AND NEW YORK

First published 2020
by Routledge
2 Park Square, Milton Park, Abingdon, Oxon OX14 4RN

and by Routledge
52 Vanderbilt Avenue, New York, NY 10017

Routledge is an imprint of the Taylor & Francis Group, an informa business

British Library Cataloguing-in-Publication Data
A catalogue record for this book is available from the British Library

Library of Congress Cataloging-in-Publication Data
Names: Jauhiainen, Jussi, 1963– author. | Hooli, Lauri, author.
Title: Innovation for development in Africa / Jussi S. Jauhiainen and Lauri Hooli.
Other titles: Rethinking development.
Description: New York : Routledge, 2020. | Series: Rethinking development | Includes bibliographical references and index. | Identifiers: LCCN 2019030653 (print) | LCCN 2019030654 (ebook) | ISBN 9780367349578 (hardback) | ISBN 9780367349561 (paperback) | ISBN 9780429328978 (ebook)
Subjects: LCSH: Economic development—Africa—Government policy. | Economic development—Africa—International cooperation. | Diffusion of innovations—Africa. | Africa—Economic conditions.
Classification: LCC HC800 .J38 2020 (print) | LCC HC800 (ebook) | DDC 338.96—dc23
LC record available at https://lccn.loc.gov/2019030653
LC ebook record available at https://lccn.loc.gov/2019030654

ISBN: 978-0-367-34957-8 (hbk)
ISBN: 978-0-367-34956-1 (pbk)
ISBN: 978-0-429-32897-8 (ebk)

Typeset in Bembo
by Apex CoVantage LLC

CONTENTS

TABLES

FIGURES

BOXES

ABBREVIATIONS

AMCOST	African Ministerial Council on Science and Technology
ARWU	Academic Ranking of World Universities
AU	African Union
BEAM	Business with Impact
BMBF	Engl. Federal Ministry of Education and Research (Germany); Deut. Bundesministerium für Bildung und Forschung
BMZ	Engl. Federal Ministry for Economic Cooperation and Development; Deut. Bundesministerium für wirtschaftliche Zusammenarbeit und Entwicklung
CADFund	China-Africa Development Fund
CII	the USAID Centre for Innovation and Impact
CNET	American media website that publishes reviews, news, articles, blogs, podcasts and videos on technology and consumer electronics globally
COFISA	Cooperation Framework on Innovation Systems between Finland and South Africa
COSTECH	Tanzania Commission for Science and Technology
CSIR	Council for Scientific and Industrial Research
CSP	Cross-sectoral partnership
DAC	Development Assistance Committee
DC	Washington, DC, formally the District of Columbia and commonly referred to as Washington or D.C., the capital of the United States

DFID	Department for International Development
DST	Department of Science and Technology (DST) of South Africa
DUI	Doing, using, interacting
EU	European Union
EUR	the euro (sign: €; code: EUR), the official currency of 19 of the 28 member states of the European Union
FDI	Foreign direct investment
GCIC	the Ghana Climate Innovation Centre
GDP	Gross domestic product
HDI	Human Development Index
HEI	Higher education institution
ICT	Information and communications technology
ICT4D	Information and communications technology for development
IMF	International Monetary Fund
ITU	International Telecommunication Union
JICA	Japan International Cooperation Agency
JKUAT	Jomo Kenyatta University of Agriculture and Technology
MDG	United Nations' Millennium Development Goals
MOST	Ministry of Science and Technology
NGO	Non-governmental organisation
OECD	the Organization for Economic Co-operation
OECD-DAC	the Organization for Economic Co-operation and Development and its Development Assistance Committee
PAUSTI	The Pan-African University Institute for Basic Sciences, Technology and Innovation
R&D	Research and development
SADC	Southern African Development Community
SDG	United Nations Sustainable Development Goals
SID	Stellenbosch Innovation District
SIDA	Swedish International Development Agency
SME	Small and medium-sized enterprises
STI	Science, technology, innovation
TANZICT	The programme Strengthening the Innovation Ecosystem in Tanzania
UN	United Nations
UNAM	University of Namibia
UNDP	United Nations Development Programme
UNECA	United Nations Economic Commission for Africa

UNICEF	United Nations International Children's Emergency Fund
US	United States
USAID	United States Agency for International Development
USD	United States dollar (sign: $; code USD) the official currency of the United States of America
UTU	University of Turku
WASH	(or Watsan, WaSH) is an acronym that stands for "water, sanitation and hygiene"

AUTHOR BIOS

Dr Jussi S. Jauhiainen is Professor of Geography at the University of Turku (Finland) and Visiting Professor of Human Geography at the University of Tartu (Estonia). His research interests are regional and urban development and policies, knowledge creation and innovation policies, as well as asylum-related migration. He has published widely on these topics in international and national journals and books. He has been Visiting Researcher at the University of Namibia and the University of Dar-es-Salaam, and has visited many other African countries for research. In Turku, he is responsible for the implementation of the memorandum of understanding between the University of Turku and the United Nations Economic Commission for Africa.

Dr Lauri Hooli is University Lecturer in Human Geography at the University of Turku (Finland). His research interests are economic geography and development studies, as well as innovation policies and practices in Africa. He has published on these topics in international and national journals and books. His PhD dissertation was about innovation policies and practices in Tanzania and Namibia, and he has been engaged with related research projects. He has lived and worked in many southern African countries.

PREFACE

This book is just a beginning. Research on innovation and development for Africa is a long journey. Innovations in Africa derive from motivated persons, communities, enterprises, universities and other actors committed to the development of Africa, its countries and people. Scholars, like ourselves, may play some role, but what is important is that Africans take part and benefit from innovations and development.

My (Jussi S. Jauhiainen) first proper trip to abroad (except as a child to the neighbouring country of Sweden and a few charter trips to Spain's sunshine) was to Morocco in the early 1980s. The popular childhood board game Afrikan Tähti (The Star of Africa), with its adventurous images, increased my curiosity to visit Africa. Morocco was the starting place of this board game. At the same time, it was the furthest place one could travel with an Inter-Rail train ticket. So, at 18 years old, I took a train from Finland to Morocco. The further south I travelled in Europe, the more I noticed differences in development. Finally, strolling along the sunny streets of Marrakesh, I got to know that like me, many young people there had dreams, ideas and ambitions. While I could realise many of my wishes, their development opportunities were constrained, ideas did not turn into innovations and many dreams did not come true.

Years later, I had visited several African countries and gotten acquainted with different people in Africa, from urban shantytowns and remote countrysides to the shiny skyscrapers and technology hubs in the capital cities. In particular, I wish to highlight the visits to Namibia in 2008, Zanzibar in 2009 and Tanzania in 2011. Each time, I could spend a few months observing the rapid everyday development that was often not so sustainable. I acknowledge the warm hospitality of the University of Namibia, the University of Dar-es-Salaam and Zanzibar University. My later visits to Africa have always confirmed that Africa is the continent of the future, but ideas need to be transformed to innovations for development.

In 2001, when I (Lauri Hooli) moved to the small mining town Selebi-Phikwe in Botswana, the expectations of African development were not that positive. The HIV epidemic was very serious, and over a third of inhabitants were affected. Botswana was peaceful, but many nearby countries still suffered from internal conflicts. On the flip side, even then you could sense the emerging mobile phone revolution. Mobile phones were even in villages without a grid. I was privileged to have half an hour of Internet time every other week. Since then, the culture of using mobile phones has become very different. The art of shortening text in messages, calling miscalls, and communal charging points have created the base for unique local need-based innovations.

In 2009, when I worked at the Embassy of Finland in Namibia, we were very excited to fund and follow the first business plan of the Namibia Innovation Centre and hear enthusiastic news about the first larger innovation-focussed development cooperation project (COFISA) from South Africa. Later, in my postdoctoral research, we followed Finnish technology companies' endeavours to get access to the technology markets of Tanzania and Kenya. The interest has been wider and faster than what I expected a few years ago. Though I left Africa many times, Africa never left me. I have a keen interest in following how my friends, colleagues and acquaintances move forward, and how cooperation between the Global North and Africa could lead to better and more inclusive innovative developments in Africa. Globally, the most exciting dynamics in innovation development are now happening in Africa – Africa is rising, and we need to support it.

We (Jussi and Lauri) never had a plan to write a book about innovation and development in Africa, at least until now, when the continent is just emerging in the innovation scene. This initial idea originated from our observations that the topic of innovation in international development aid and cooperation has barely been addressed in research, especially regarding Africa. In addition, we have been surprised by how seldom the key policy-makers and other actors scrutinise or conceptualise what innovations actually are, what impact they have, and what could be Africa's advantages and challenges.

Looking backward, we realise that we had touched this topic in many of our articles. Nevertheless, when the idea unintentionally sparked in a conference, and Dr Helena Hurd and Leila Walker from Routledge/Taylor & Francis took a supportive stance on it, then it was just a matter of time to submit the manuscript. Probably the deadline (set by us ourselves!) was so tight that we never had second thoughts about such an ambitious project. The editorial and technical staff at Routledge made our path from a vague idea to a proper book as smooth as possible. The remaining errors and potential misunderstandings in the book are ours. However, this book is not our final word, but more like opening phrases about the topic.

We thank the Division of Geography at the University of Turku, who gave us some time to write the manuscript. We are not native English speakers, so the Division's contribution to the language checking costs is much appreciated, as well as the work of anonymous proof-readers somewhere in the spheres of the Internet. Interest in Africa is on the rise at our university, and for this we express our gratitude to the University of Turku former Rector Kalervo Väänänen and the current Rector Jukka Kola for their encouragement and support regarding Africa-related activities. Our colleagues at the department and the KREPRO (Knowledge Creation Processes) research group have always been supportive. In particular, we thank Petteri Savolainen for helping us with the design of the figures. We did not get any dedicated funding to write this book. However, the project URMI (nr 303617) by the Strategic Research Council at the Academy of Finland, the PUT project (PRG306) at the University of Tartu and the Business Finland BEAM project gave us opportunities to be engaged with our research.

Writing this book was a short but intensive exercise. The people near to us felt that we spent too much time behind our laptops. Now, finished with this exercise, we wish to return to spending time with our families, who deserve more of our attention than articles, books, laptops and the eternal depths of the Internet.

Turku and Tartu, June 2019
Jussi S. Jauhiainen & Lauri Hooli

1

INTRODUCTION TO INNOVATION FOR DEVELOPMENT IN AFRICA

Introduction

Africa is undergoing a profound transformation that is changing the lives of hundreds of millions of people on the continent. Rapid urbanisation, demographic growth and migration, technological development and digitalisation, as well as globalisation and structural economic changes, will be key issues in Africa in the 2020s and 2030s. These will result in enormous local and global changes. The continent and the world will not be the same, and Africa's resilience is crucial to the future of the planet.

Some ongoing changes, such as economic growth and new technologies, offer major opportunities for the continent's prosperous future. Africa also faces significant challenges, such as harmful global environmental and climate change, rapid population growth leading to potentially uncontrollable urbanisation, and increasing inequality resulting from rapid economic growth, all of which may further marginalise the continent and many of its inhabitants.

Africa's future resilience depends on the continent's, countries', and inhabitants' ability to harness innovation and create innovative development systems. For decades, well-functioning innovation systems have been regarded as major economic growth and job creation stimuli in the Global North (i.e. in the most developed parts of Europe, North America, Asia and Oceania). Innovation also supports positive socio-economic transformation there. In the Global South (i.e. in Africa, Latin America and developing Asia, including the Middle East) innovation and innovation systems must be at the forefront of development strategies, policies, and practices to reduce the poverty and inequality that many Africans face. However, as Schillo and Robinson (2017) note, innovations can also contribute to increasing economic and social inequalities.

It is impossible to discuss Africa in a monolithic sense because it is so ethnically and culturally diverse. Africa consists of 54 independent countries of varying sizes, populations, and geographic locations. The largest country, Nigeria, has over 200 million inhabitants, whereas the smallest countries have only a few hundred thousand. Africa's population is growing rapidly. Of its 1.3 billion people, the majority are young adults or children. The median age of Africans (19.4 years) is less than half of that of Europeans (42.9 years) (United Nations, 2019a). Northern African countries are relatively well developed, but Sub-Saharan Africa remains, with the exception of South Africa, the poorest region worldwide. In addition, in 2015–2018, the annual economic growth in Sub-Saharan Africa remained below 3% (World Bank, 2019).

Thousands of distinct languages and cultures coexist in Africa. Despite recent projects to create an intra-African free trade area and pan-African development policies, there is little policy harmonisation among African countries. Even the continent's five macro-regions exhibit vast internal differences (Figure 1.1).

Innovation for Development in Africa analyses and observes innovation and development in Africa, and in particular how international development policies, cooperation and innovation meet and interact in Africa. Innovations – new or improved products, services, and processes – are the main objectives of international aid to Africa, its countries and regions, and the developing world in general. This book addresses two closely interconnected socio-economic dynamics – innovation development and international development aid – that play a major role in current and future socio-economic developments in Africa, as well as the development nexus between the Global North and the Global South.

First, the book contextualises the rapid development of knowledge-based societies, in which knowledge is transformed into innovations, and discusses this model's implications for Africa. The creation, diffusion and use of knowledge and innovations are considered fundamental to beneficial socio-economic transformation and sustainable economic growth. The socio-economic success of countries in the Global North – Europe, North America, and Asia – derives from their public, private and non-governmental sectors' capacity to create added value through innovations supported by well-established innovation systems, which in turn are set and supported by

FIGURE 1.1   Countries and macro-regions of Africa

public development policies. In this sense, innovation systems connect local and global economies and people.

Second, the international development (aid) regime is in transition due to changes in relations between the Global North and South. Previously, the divide between the substantially wealthier and more developed Global North and the less developed Global South was clear. However, in recent years, rapid economic growth in the Global South has shifted many previously less-developed countries to transitioning and middle-income countries. Such changes have ruptured the customary global development axis. For example, South-South partnerships are becoming increasingly important, including the substantial role of China and Chinese investors and economic actors. Their cooperation follows only partly, if at all, the logic of the development assistance overseen by the Organization for Economic Co-operation and Development and its Development Assistance Committee (OECD-DAC). Nevertheless, North-South and South-South development cooperation innovations have gained significance.

Political changes, many following the global economic crisis of 2007–2008, transformed development ideology in the Global North, and the so-called beyond-aid agenda emerged. Aid is no longer considered a one-way donation from wealthier countries to developing countries. Instead, it is based on expectations of mutual economic benefits and reciprocity between donors and recipients. Development assistance is increasingly linked to other aspects of international politics, such as foreign trade, global environmental protection, and technological development. Behind this new development orientation are also political changes, especially the rise of new nationalist and populist political movements in the Global North. These movements advocate withdrawal from free global trade and macro-regional development regimes, and support a return to nation-states that exercise individual privileges. Proponents of this model claim it is necessary to care for the citizens of one's own country before supporting the poor in other countries.

This political development has changed development aid. Development assistance budgets have increased worldwide. Between 2000 and 2017, the amount of net official development assistance and official aid received tripled to $163 billion USD (OECD, 2019). However, a substantial part of these funds is redirected to serve donor countries' endeavours. For example, donor countries' private-sector activities, exports and foreign trade (Murray & Overton, 2016) are supported in countries that receive aid. In addition, development funds are blurred into broader international politics, such as mitigating the costs of refugees arriving in the Global North – which proponents eagerly label "the refugee crisis". Along with this development, new global actors have emerged in the development field. Some have been initiated by wealthy individuals and the foundations of large multinational corporations. These intervene in development in the Global South without consent or even connection to any government, whether in the Global North or the Global South. The wealthiest of these is the Bill & Melinda Gates Foundation, which donates billions of USD annually to improve health and development in developing countries and to foster their social transformation (Ratha et al., 2008; Levich, 2015).

Innovation and support for innovation development have become primary international development policy and practice objectives in Africa. Their aim is to enhance economic growth and employment throughout the continent, alleviate poverty and reduce inequality. The most economically significant donor countries – the United States (USAID, 2019), the United Kingdom (DFID, 2017) and Germany (BMZ, 2017) – as well as international organisations, such as the United Nations Economic Commission for Africa (UNECA), the United Nations Children's Fund (UNICEF), the Organization for Economic Cooperation and Development (OECD), the World Bank, the Southern African Development Community (SADC) and various European Union (EU) development initiatives for Africa, increasingly mention innovation as their key development objective in Africa. In addition, several other global, national, and local development agents prioritise innovation in their African agendas. In particular, China, the most significant international investor in Africa, connects innovation-related issues to its major ongoing structural development processes and projects on the continent (Brooks, 2019). It has been estimated that China by the year 2018 financed $57.5 billion USD of infrastructure development in Africa. It is financing one out of six, and constructing one out of three, infrastructure projects in Africa (Deloitte Africa, 2018).

Innovation plays a central role in several important SDGs launched by the UN (2015), especially SDG 9 (*Build resilient infrastructure, promote inclusive and sustainable industrialisation and foster innovation*). The AU Agenda 2063 emphasises making science, technology and innovation the main engines of socioeconomic development in Sub-Saharan Africa and in its plan for transforming Africa into the global powerhouse of the future (African Union, 2015). In recent years, many African countries have launched policies and projects to promote innovation. A comprehensive approach to systematic innovation development has become policy in many African countries, including Algeria, Cameroon, Côte d'Ivoire, Egypt, Ethiopia, Ghana, Kenya, Malawi, Morocco, Namibia, Nigeria, Rwanda, South Africa, Tanzania and Zambia, as well as their international development cooperation (see Lemarchand and Tash, 2015; SIDA, 2015; Tigabu et al., 2015; Hooli et al., 2016; Moon et al., 2016; Jauhiainen & Hooli, 2017; Hooli et al., 2019). Innovation will likely soon be mentioned in almost all African countries' development strategies.

Changing development dynamics in Africa – Africa rising?

During the early decades of the 21st century, Africa has become the continent of high hopes and untapped opportunities. Many countries have witnessed extremely rapid economic growth. Several of the fastest-growing economies worldwide are in Africa (World Bank, 2019). Most of the continent's countries are liberalising their trade policies and are increasingly interconnected to the world economy (Shizha & Diallo, 2016). In 2018, the African Continental Free Trade Agreement created the world's largest free-trade zone to facilitate business growth and intra-African trade. The evidence and future predictions of such positive developments have increased interest and investment in African resources and talents.

So far, much African economic growth is based on expanding global demand for raw materials and natural resource exploitation. This is supported by massive population growth, increasing continent-wide urbanisation and wide distribution of information and communication technologies (ICTs) that connect even non-electrified African rural villages to global information flows and markets. Smart phones are present almost everywhere in Africa, and many who were unable to afford earlier expensive broadband connections can now access the Internet for the first time. Nevertheless, in 2016, countries where the majority of population used the Internet were only Morocco, Seychelles, South Africa, Mauritius and Tunisia. Less than 10% of the population used the Internet in 14 African countries (UNECA, 2018a: 108). The situation is changing fast, and the association of mobile operators GSMA (2018) estimates that by 2025, Africa will have 300 million new Internet users, doubling the number of users in Africa in 2018 (ITU, 2019). Although agricultural productivity remains quite low, there are high hopes the Green Revolution will soon reach Africa. New crops and agricultural mechanisation would increase the continent's food security and make it the global breadbasket as its agricultural technology advances (Mallory & Giuliani, 2018; see also Holt-Gimenez et al., 2008). Approximately 60% of the world's uncultivated arable land is in Africa (African Development Bank, 2018), which makes agriculture a potentially significant activity in Africa in the future. Today, in many African countries the majority of the population is still employed in agriculture, and many in economically inefficient small-scale agriculture.

Nevertheless, despite rapid economic growth and increasingly peaceful development, the continent's most burning development problems remain unsolved because its economic development has been neither inclusive nor pro-poor. In 2015, 41% of the population in Sub-Saharan Africa still lived below the poverty line – i.e. 413 million people there lived on less than $1.90 USD per day (Atamanov et al., 2019). The majority (56%) of the global poor live in Sub-Saharan Africa, and this population is expected to grow. The continent's people have not significantly benefitted from the recent economic growth. Economic inequality in Africa continues to increase (Odusola et al., 2017). More than a third of adults cannot read and write, and many indigenous groups are increasingly marginalised.

However, the continent's demography leads to situations in which hundreds of millions of people are expected to enter the job market during the next few years. When people do not perceive suitable living conditions in their home countries, they migrate. Between 1980 and 2010, the number of African migrants doubled to 30.6 million, and the proportion of migrants leaving Africa rose from 41% (6 million people) to 49% (15 million people; Ehrhart et al., 2014). Migration to the Global North has intensified in the 2010s. This has removed the most talented individuals from already talent-scarce pools and encouraged desperate, uneducated youths to sell their worldly goods to human traffickers and risk their lives to cross the Sahara and the Mediterranean Sea to seek better opportunities in Europe (see Jauhiainen, 2017a, 2017b). If they make their way to Europe and find employment, many send later remittances to their families remaining in Africa, which is a substantial help both for the families and the national economies (Ratha et al., 2008). As part of Africa's intensified connection to the global economy, its abundant raw materials are exploited and exported without processing them on the continent, and added value is generated elsewhere. As a result of such exploitation, urbanisation and population growth, the environment is worsening, and the harmful effects of global climate change have become increasingly evident. This will increase migration flows in the future.

Good governance and democracy form one cornerstone of development. In general, the number of wars has declined worldwide in the 2010s, but a few major wars and terrorist groups persist in many places. In 2018–2019, wars in eight African countries resulted in over 1,000 casualties, and in one-third of African countries, conflict-related fatalities numbered between 100 and 1000. However, the majority of African countries have experienced no wars for the past two decades. During the first two decades of the 21st century, political leadership has changed in 19 African countries due to revolution. In 2019, still 10 out of 54 African presidents and other national leaders had been in power for more than two decades. Many presidents and prime ministers have changed in the 2010s. The long-lasting authoritarian leaders are gradually disappearing from the political landscape, as the recent examples of Zimbabwe, Algeria and Sudan indicate. These leaders were toppled after months of popular protests. Unfortunately, political changes that overthrow dictators or authoritarian regimes often usher in new corrupt regimes, and many citizens have lost faith in democratic leadership and good governance in Africa.

Judd Devermont, the director of the Africa Program at the Center for Strategic and International Studies, notes that in 2008, Sub-Saharan Africa experienced around 800 protests, whereas during 2018, the number of protests was just below 4,000 (CNET, 2019). Social media and ICTs have begun to play important roles in Africa's democratisation. Citizens use mobile phones and social media to organise and inform the masses and visualise protests locally, nationally and globally. In 2019, Sudan's popular revolution was iconised by one young woman in white clothing standing on top of a car to address protesters around her – protesters with mobile phones in their hands. Some authoritarian countries in Africa have even attempted – unsuccessfully – to restrict or eliminate access to the Internet.

Nevertheless, the political changes of the 2010s have not yet resulted in democracy in many African countries (Figure 1.2). The political economy of the continent is very complex (Amin, 2014). On the global Democracy Index, of the 167 countries listed, the three highest-ranked African countries – with the exception of the small island states of Mauritius and Cabo Verde – are in positions 28 (Botswana), 40 (South Africa) and 56 (Lesotho). Of the 20 least democratic countries, half are in Africa – and the Democratic Republic of the Congo is at the bottom (Economist Intelligence Unit, 2019).

In economic development, other cornerstones, recent relative GDP and productivity growth, have been rapid, as mentioned previously. Africa is emerging as a global economic player, if considered collectively. In 2018, eight African countries were among the 20 fastest-growing economies worldwide, and Libya, Rwanda, Ethiopia and Côte d'Ivoire were among the top five countries with real annual GDP growth of over 7% (IMF, 2019). Several of these countries had experienced such growth for years. In the coming decades, several features will support strong economic growth. Among these are increased global demand for raw materials, rapid urbanisation and related economic development in Africa, enhancement of general education levels in Africa, and potential gradual eradication of poverty in Sub-Saharan Africa. With population growth and increased general wealth, Africa will become the next major market for consumer goods, some of which are made locally, but most of which originate outside Africa. Nevertheless, half of the African countries still have low income levels (Figure 1.3).

When studying the continent's impressive relative economic growth, one should remember that the initial figures for most African countries are small. In absolute terms, most African countries still have small economies. In 2018, Nigeria's and Egypt's economies, the largest on the continent, were the size of Poland's. The 20 smallest African economies together are the

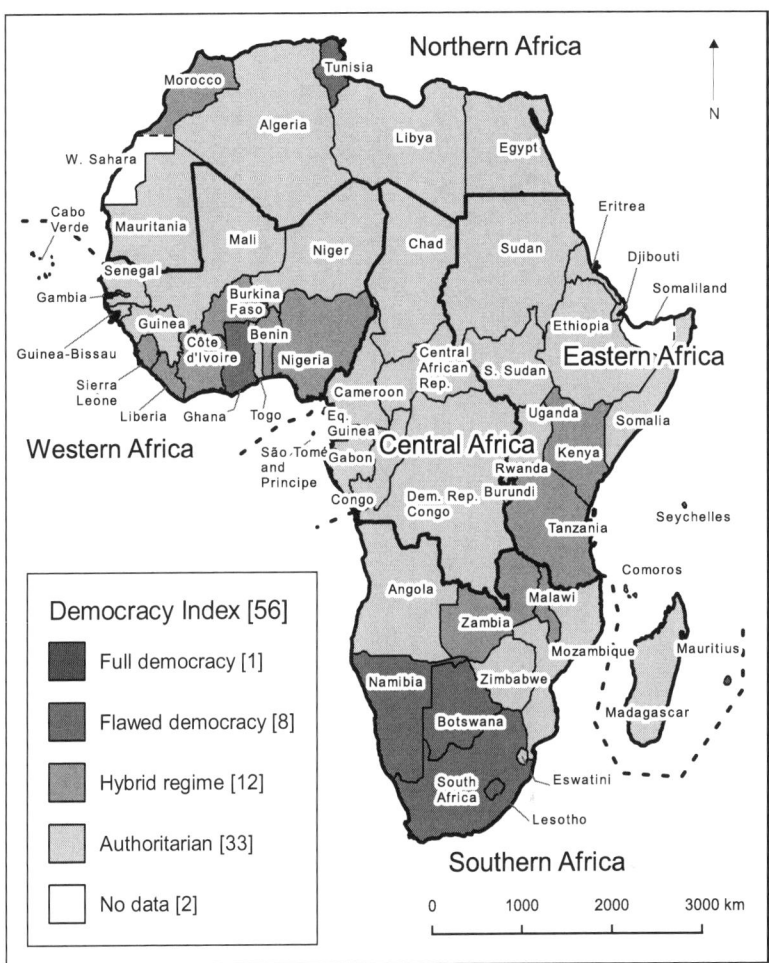

FIGURE 1.2   Politics in Africa

Source: Modified Economist Intelligence Unit (2019)

size of New Zealand's economy. Therefore, despite rapid economic development, most African countries have not yet transformed their potential into large GDPs due to their less-developed economic structure. In 2018, agriculture accounted for 55% of Sub-Saharan Africa's active workforce, and in 18 African countries, over two-thirds of the active labour force were employed in agriculture (UNECA, 2018b). Furthermore, agricultural productivity and wealth creation are generally quite low. Currently, 85% of

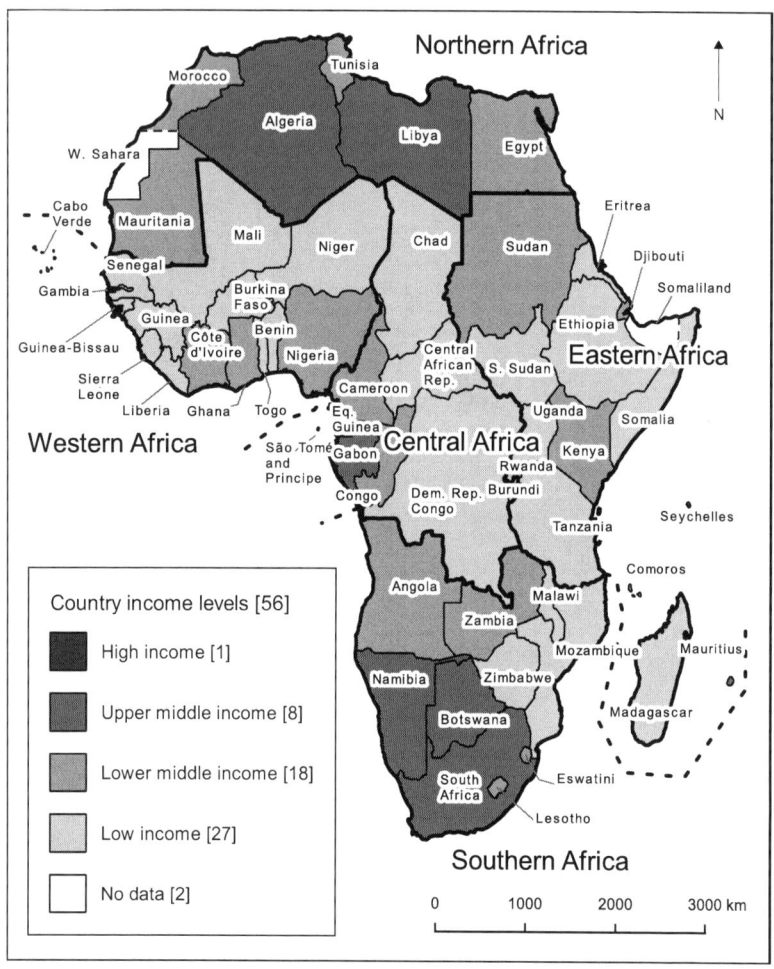

FIGURE 1.3 Economy in Africa

Source: Modified from UNECA (2018b)

African farms occupy less than two hectares. Furthermore, global climate change threatens these fragile fields. In addition, a major complexity of the African economy derives from the fact that much of it is informal. According to Galdino et al. (2018), the informal economy accounts for up to 90% of jobs in the lowest-income Sub-Saharan African countries, such as the Central African Republic and the Democratic Republic of the

Congo. However, the informal economy and growing population also provide opportunities for need-based innovation.

Supported by a large population, population growth, purchasing power and digital economies, some African enterprises have begun to grow into big businesses. South African companies dominate the rankings. In 2018, of the 30 largest companies in Africa, 26 were from South Africa, as were 67 of the 100 largest enterprises. In 2014, more than 400 enterprises had annual revenues of more than $1 billion USD, and 13 companies had annual revenues of more than $10 billion USD. In total, the 700 largest companies earned a combined $1.4 trillion USD. However, the geography of big enterprises (those with turnovers of more than $500 million USD) is uneven. Approximately half (49%; 300 such enterprises) were located in South Africa alone, and a total of 56% were located in Southern Africa. Just over one-fifth (22%) of all large enterprises were in Northern Africa, mostly in Morocco, Algeria and Egypt, in the Mediterranean region close to Europe. One in seven (15%) big enterprises were located in Western Africa, and the majority were in Nigeria (56 such enterprises). Although East Africa has often been the focus of emerging enterprises, its share of large enterprises in Africa was only 4%, followed by the generally less-developed Central Africa, which had 3% of all enterprises with revenues more than $500 million USD annually (Leke et al., 2018). However, in 2019, no Africa-based enterprises appeared on the global Fortune 500 list, which includes the largest enterprises in the world (Fortune, 2019).

The third cornerstone of innovation development in Africa is its population's skills. In many countries, less than 10% of the youth attend tertiary education (Figure 1.4). Nevertheless, the number of Africans in higher education has risen quickly and is expected to grow. New public and private universities have been established to meet increasing demands for higher education, but most African universities focus on basic undergraduate education (see Chapter 7). The expansion of higher education has also meant that not all universities meet international performance standards. Compared with universities in the Global North, African universities focus substantially less on both basic and applied research. The number of star scientists – leading international scholars – is rather small in Africa and non-existent in many African countries (Doh & Jauhiainen, 2019). According to the global ranking of universities (ARWU, 2018), five African universities feature in the top 500 universities globally. Four of them are in South Africa (University of the Cape Town is ranked highest in positions 201–300), and one is in Egypt. In positions 501–1000, there are 10 more African universities. The top 10 and top 50 African universities are presented in Figure 1.4. However, the overall lack of advanced R&D competences hinders the establishment of a proper

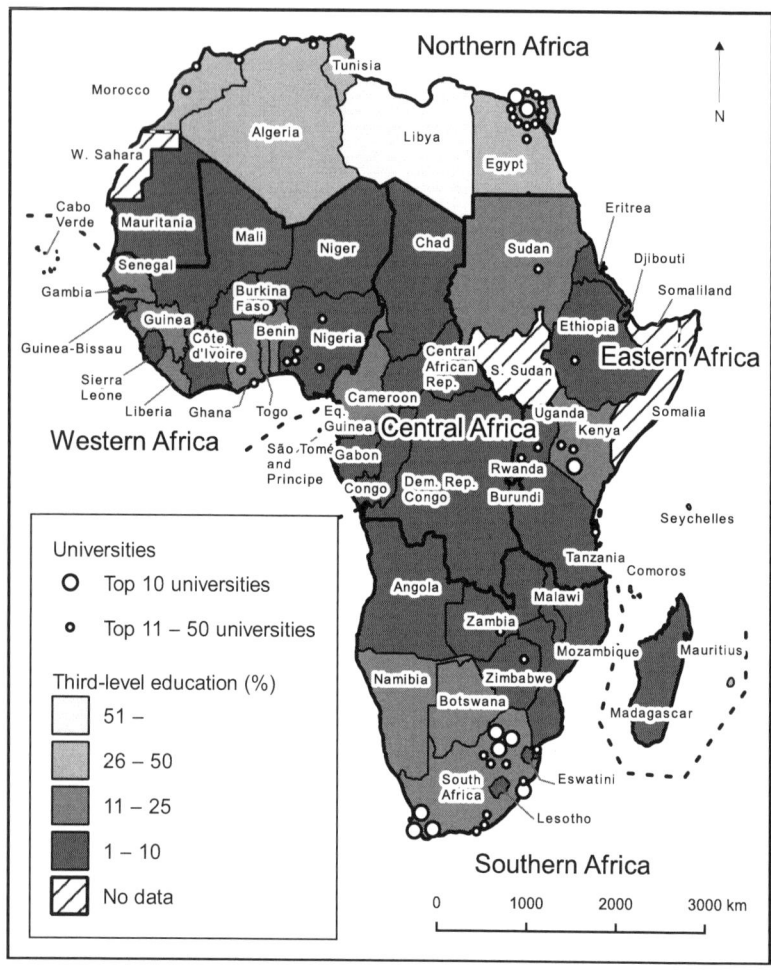

FIGURE 1.4 Skills in Africa

Source: Modified from UNECA (2018b), ARWU (2018) and Cybermetrics Lab (2019)

impactful innovation systems in African countries other than South Africa and, to a lesser degree, Egypt (see Chapters 2 and 4).

Due to African universities' limited resources, many skilled people and capable scholars must leave Africa to advance in their academic fields that require excellent laboratories and research teams. This brain drain is a development issue in Africa, but it could become an opportunity if trained, skilled expatriates return. In addition, hundreds of thousands of African graduates and post-graduates search for opportunities to apply their knowledge and skills for the good of society. The wide variety of languages and cultures in Africa also creates potential for innovation. Thousands of diverse socio-economic cultures possess indigenous knowledge that could be the source of innovations that cannot emerge elsewhere (see Chapter 8). However, suitable paths must be established to convert indigenous knowledge into global commercial innovation, and these paths are not always obvious (Jauhiainen & Hooli, 2017; Hooli et al., 2019).

The fourth cornerstone of development, urbanisation, is linked to all previously mentioned development aspects. In general, until recently, a model has prevailed which states that the more urbanised a country is, the higher its GDP per capita is. Rising urbanisation is connected to structural economic shifts toward increased employment in the service industry and skilled, information-based fields. Rising urbanisation also opens new market niches in various consumption fields.

Increasing urbanisation drives and will drive Africa's future. In the past two decades, Africa's urban population has grown to match that of the EU. Urban growth in Africa is also accelerating, and the urban population there is expected grow from 550 million people in 2018 to 824 million by 2030, 1.125 billion by 2040, and 1.489 billion by 2050 (United Nations, 2019b). Along with this general urbanisation, the number of cities with more than one million inhabitants is increasing rapidly. In 2019, approximately 50 African cities had at least one million inhabitants, and four cities had over five million inhabitants. Some metropolitan areas, such as Lagos in Nigeria (21 million inhabitants), Cairo in Egypt (18 million) and Kinshasa in the Democratic Republic of the Congo (13 million), have become megacities.

Urbanisation is a challenge and an opportunity. It requires a vast amount of infrastructure, but many jobs will be created and increase demand for local and non-local products. New urban residents require inclusive innovations to improve their lives and livelihoods. In fact, Nawrot et al. (2018) illustrate how African megacities as dynamic innovation ecosystems can be drivers of the continent's economic transformation. However, better forms of governance and planning, as well as inclusive economic growth with a more sustainable society, are prerequisites for a rising Africa and its (mega)cities.

Purpose and outline of this book

Innovation is discussed, applied and developed almost everywhere. However, scholars have not yet properly addressed the intertwining of innovation, innovation policies and international development policies. State-of-the-art

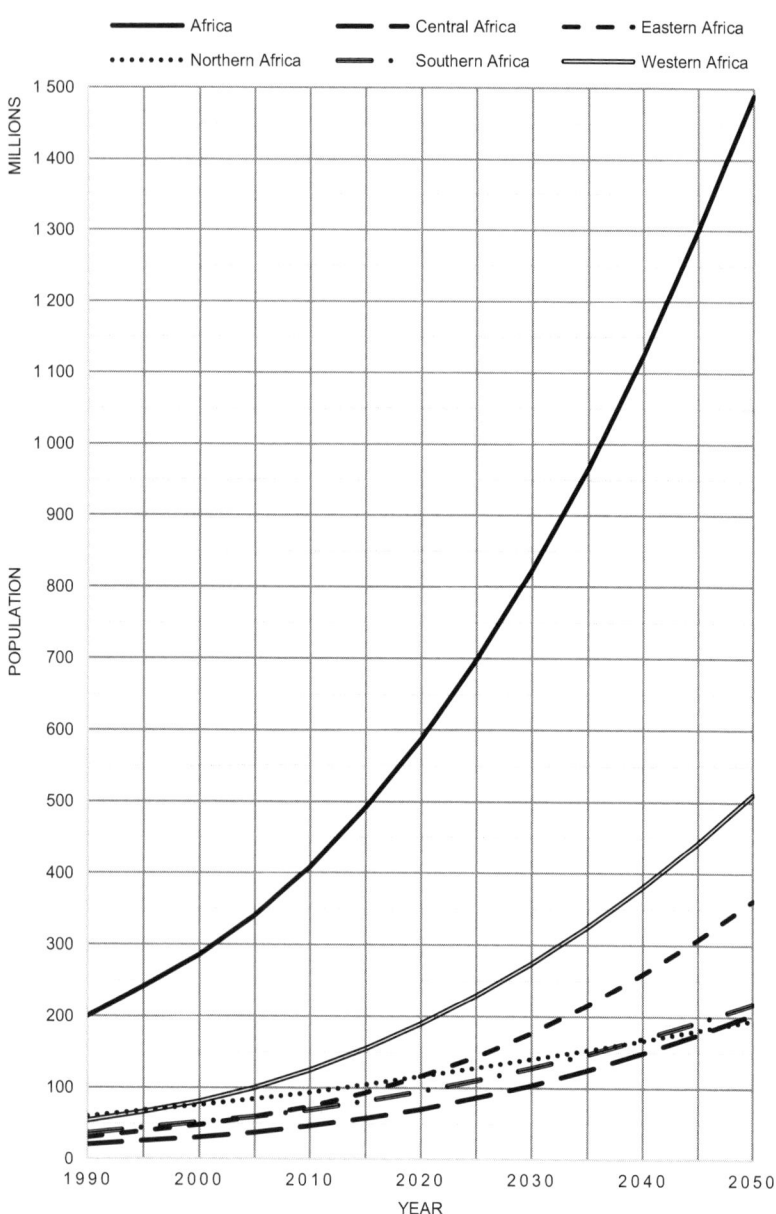

FIGURE 1.5 Urbanisation in Africa

Source: Modified from the United Nations (2019b)

knowledge about the rationales, practices and outcomes of innovation-focussed development and policies in African countries is needed. When knowledge, competence building and innovation are raised as a topic, there is an overlying ambiguity and fuzziness regarding what innovation means. Development policies and related academic literature usually take innovation for granted without examining the conceptual and contextual understanding of its specificities in Africa.

Innovation for Development in Africa provides a research-based analysis and overview on how international development policies, cooperation, and innovation meet and interact in Africa. Among its key themes are changing development policies between the Global North and South; the role of innovation in international aid and development policies; private, public, non-government, university and other development actors; and the potential for inclusive innovation in local communities.

This book deals with two closely interconnected socio-economic dynamics – innovation development and international development aid – that play major roles in current and future African socio-economic developments and in the development nexus between the Global North and South. Research suggests a positive connection between innovations, urbanisation and regional economic growth. However, the connection between general development objectives, innovation promotion and the outcomes of their intertwining in developing countries is much more complex. This book derives from the authors' research into innovation and international development policies in Africa, the most recent scientific research, and contemporary development policies and practises.

This book originated from our observations that the topic of innovation in international development aid and cooperation has not been addressed in research, especially regarding Africa. During our field trips and study periods in Africa over the past two decades, we have observed how innovations are changing the continent's development scene. On the one hand, creativity and the need for novelty are present in many Africans' everyday lives, whether they struggle to survive on the margins or set the scene for globally relevant mobile applications. On the other hand, innovations are present in international development aid strategies, policies and projects for African countries.

When engaging with public development aid donors and recipients, we have been surprised by how little attention key stakeholders pay to innovation and how seldom they scrutinise or conceptualise what innovations actually are, what impact they have, and what could be Africa's advantages and challenges. There have been high hopes of rapid, direct policy and technology transfers from the Global North to the Global South without proper consideration of risks and failures. Although public development aid projects have at the same time addressed development with budgets counted in millions of USD, private sector actors have invested and transformed Africa with billions of USD. Such perplexity encouraged us to study these topics both independently and with African and European colleagues.

In the background, we elaborated on the concept of knowledge and how knowledge develops through different spatio-temporal contexts into innovation (Hautala & Jauhiainen, 2014). The mainstream innovation research focusses mostly on large urban settings and high technology in the developed Global North. Such studies seldom fit the African context, even for large, rapidly growing cities, such as Lagos in Nigeria or Dar-es-Salaam in Tanzania. Therefore, we emphasise the study of the development of knowledge, innovation, innovation systems, and contemporary features of innovation development, such as living labs in peripheral settings in remote areas in Europe and Africa (Jauhiainen, 2006; Jauhiainen & Suorsa, 2008; Jauhiainen & Moilanen, 2012; Hooli et al., 2016; Hautala & Jauhiainen, 2019). In Africa's changing environments, partly due to global climate change, indigenous knowledge mixes with modern science-based knowledge, for example, in local communities' attempts to understand changing weather and climate patterns to develop more resilient local ways of living (Hooli, 2016).

In the end, academic texts about knowledge and innovation easily distance themselves from the everyday realms in which policies are created. Innovation systems have become a practical framework for many African countries to promote development innovation, at least on paper, and many studies mention innovation systems in Africa, but few analyse their development. This motivated us to conduct a detailed and precise longitudinal analysis of the development of national innovation systems in Namibia. This case has also provided also opportunities to address the potential benefits and pitfalls of indigenous knowledge as part of innovation systems (Hooli & Jauhiainen, 2017; Jauhiainen & Hooli, 2017). As we mentioned previously, indigenous knowledge offers a potential competitive advantage in an IS. Nevertheless, the complexity of commercialising indigenous knowledge is also present in African innovations, including those related to tourism (Jauhiainen & Hooli, 2017; Hooli, 2018).

Finally, we recognise that innovations are becoming key tools within the context of international development aid to Africa, a trend that indicates a broad change in the Global North's development policies (Hooli & Jauhiainen, 2017; Hooli & Haaranen, 2019). Innovation can become a transformative engine for change in Africa. Novel innovation policies in Africa can become open-ended, diverse and experimental to be more socially inclusive and environmentally sustainable (Hooli et al., 2019).

Chapter 1, as indicated previously, presents the context of this book: the emergence of innovation and development policies focussing on innovation. We highlight how innovation has become central to global development, and discuss the opportunities and challenges the trend affords Africa. Africa is the only continent on which population growth continues. In addition, the continent is undergoing intense urbanisation that will result in over one billion new urban dwellers in just a few decades. This is a significant opportunity to grow away from poverty, but innovations are necessary to achieve this goal. In the last part of this chapter, we present the book's structure and explain how its topics address various aspects of international development cooperation, focussing on innovation.

In Chapter 2, we raise the issue of innovation in the African context. We begin by illustrating the general consensus that knowledge and innovation are vital to attaining national socio-economic prosperity. Thereafter, we present an overview of the foundations of innovation and show how innovation development has evolved since World War II from linear innovation models toward a comprehensive innovation system approach, as well as recent demands for transformative innovation policies. We focus specifically on Africa, elaborate on the challenges of innovation as they relate to reducing inequality and poverty, discuss the knowledge-based approach to innovation, and suggest foci for transformative innovation policies in African countries.

In Chapter 3, we synthesise the trajectory of international development aid and cooperation in recent decades. We also highlight the main arguments for and against various approaches to assisting developing countries. Such approaches include modernisation, structural adjustments through the Washington consensus, the post-Washington consensus focus on human and sustainable development through Millennium Development Goals and SDGs, and the so-called beyond aid agenda that seeks reciprocal economic benefits for aid donors and recipients.

In Chapter 4, we explain the current understanding of innovation-based development aid and its related policies, projects and practices. Such development policies are designed by donor governments in the Global North and implemented by the private sector enterprises and entrepreneurs from the donor and recipient countries. Such policies and practices highlight the role of science, technology, and innovation in knowledge and innovation development. However, there is increasing interest in bottom-up approaches that use inclusive innovations to support indigenous people and their knowledge, thus providing more context-specific opportunities for poverty eradication.

Thereafter, in Chapters 5–8, we focus on various key actors in innovation-focussed development. In Chapter 5, we present the private sector's manifold and complex role in innovation-based international development cooperation. The private sector has become a major actor in contemporary development cooperation. Donor countries benefit from their donations by supporting their domestic companies' attempts to gain new markets and customers in the Global South. In addition, current development policies and practices strengthen and build the capacity of Africa's emerging local private sector, particularly the local technology start-up scene. There are also attempts to engage with the continent's vast informal sector and convert it into a local development trigger.

In Chapter 6, we discuss the role of African universities in knowledge creation, innovation systems and development policies. African countries host well over 1,000 universities, but only five of them are among the top 500 universities worldwide. These institutions are diverse: many focus on basic education for large undergraduate populations, whereas others focus on the applied technical sphere. Few focus on international science and high-level academic research. In general, universities that generate advanced knowledge are key actors in innovation systems, and their role is recognised in various knowledge-based development models. International development aid helps select universities to build capacity. In seeking more relevance in African society, some African universities try to be entrepreneurial.

In Chapter 7, we discuss a variety of actors who do not directly belong to the public or private sector, such as non-governmental organisations, local development groups, global foundations and activists dedicated to fostering innovation and innovation-driven development in the Global South. Sometimes the actors focus on tangible local issues, but at other times, their scope is truly global. For example, with their support, advanced technologies changed the lives of hundreds of millions of people in the Global South. These actors are seldom publicly accountable for their development strategies, and their connections to public policies or private sector activities vary considerably.

In Chapter 8, we focus on the role of local inhabitants and communities in innovation-focussed development and projects that emerge from related development aid. For long-term, sustainable impact, local communities should actively participate in development and projects aimed at transforming the society and environment around them. International development aid supports individuals and communities in the Global South. Although earlier support was often top-down and directed from the Global North, interaction between aid donors and recipients is now sought. However, new policies and practices are often directed at entrepreneurs who must take responsibility for themselves, take advantage of innovation-related processes, and produce innovations through international development aid. This is a key concern related to innovation-focussed, beyond-aid development, because not all impoverished Africans desire to become – or can become – entrepreneurs. Instead, opportunities need to be sought from inclusive innovations, sometimes supplied with indigenous knowledge. Inclusive living labs are one instrument for such opportunities.

Finally, Chapter 9 is devoted to conclusions derived from the topics discussed throughout the book. We also present recommendations regarding international development aid policies and innovation-focussed practices, and suggest topics for future research. Finally, we welcome all researchers to study these topics, reflect, and criticise our findings to advance the research frontier and achieve long-term, sustainable development in Africa and beyond.

- The development patterns related to aid and cooperation between the Global North and South are changing.
- Innovations that generate economic and social transformations are becoming the main objective of international development aid to Africa and the developing world.
- Innovation-focussed development aid faces challenges in bringing equality to Africa and reducing poverty there.

Discussion questions

- What are the major changes in development cooperation between the Global North and the Global South?
- What opportunities and challenges exist for a positive transformation of Africa?
- Discuss what you know and what you do not know about Africa.

References

African Development Bank (2018). *Agri-Tech Can Turn African Savannah into Global Food Basket*. www.afdb.org/en/news-and-events/agri-tech-can-turn-african-savannah-into-global-food-basket-african-development-bank-18609/. Retrieved May 2019.

African Union (2015). *Agenda 2063: The Africa We Want*. www.au.int/. Retrieved June 2019.

Amin, S. (2014). Understanding the political economy of contemporary Africa. *Africa Development* 39:1, 15–36.

ARWU (2018). *Academic Ranking of World Universities*. www.shanghairanking.com/ARWU2018.html.

Atamanov, A., Castaneda Aguilar, S., Corral Rodas, P., Dewina, R., Diaz-Bonilla, C., Jolliffe, D., Lakner, C., Lee, K., Montes, J., Moreno Herrera, L., Mungai, R., Newhouse, D., Nguyen, M., Beer Prydz, E., Sangraula, P. and Yang, Y. (2018). *Global Poverty Monitoring Technical Note*. March 2019 PovcalNet Update.

BMZ (The Federal Ministry for Economic Cooperation and Development) (2017). *Harnessing the Digital Revolution for Sustainable Development. The Digital Agenda of the BMZ.* www.bmz.de/. Retrieved June 2019.

Brooks, L. (2019). The 21st century belongs to China – but the 22nd will be Africa's. *Quartz*, February 21.

CNET (2019). *The Internet Is Changing Africa, Mostly for Better.* www.cnet.com/news/the-internet-is-changing-africa-for-the-better-chan-zuckerberg-gates/. Retrieved May 2019.

Cybermetrics Lab (2019). *Ranking Web of Universities.* www.webometrics.info/. Retrieved June 2019.

Deloitte Africa (2018). *Africa Construction Trends Report 2018.* Deloitte.

DFID (Department for International Development) (2017). *Economic Development Strategy: Prosperity, Poverty and Meeting Global Challenges.* DFID, London.

Doh, P. and Jauhiainen, J. (2019). A multi-level analysis of the success factors of aspiring African star scientists in the international scientific networks. Submitted manuscript.

Economist Intelligence Unit (2019). *EIU Democracy Index 2018 – World Democracy Report.* www.eiu.com/. Retrieved May 2019.

Ehrhart, H., Le Goff, M., Rocher, E. and Singh, R. (2014). Does migration foster exports? Evidence from Africa. *World Bank Policy Research Working Paper* 6739. World Bank, Washington, DC.

Fortune (2019). *Fortune 500.* www.fortune.com/fortune500/list/. Retrieved June 2019.

Galdino, K., Kiggundu, M., Jones, C. and Ro, S. (2018). The informal economy in pan-Africa: Review of the literature, themes, questions, and directions for management research. *African Journal of Management* 4:3, 225–258.

GSMA (2018). *The Mobile Economy: Sub-Saharan Africa 2018.* GSMA.

Hautala, J. and Jauhiainen, J. (2014). Spatio-temporal aspects of knowledge creation. *Research Policy* 43, 655–668.

Hautala, J. and Jauhiainen, J. (2019). Creativity-related mobilities of peripherally located artists and scientists. *Geojournal* 84:2, 381–394.

Hooli, L. (2016). Resilience of the poorest: Coping strategies and indigenous knowledge to live with the floods in Northern Namibia. *Regional Environmental Change* 16:3, 695–707.

Hooli, L. (2018). From warrior to beach-boy: Resilience of Maasai in Tanzanian tourism industry. In Cheer, J. and Lew, A. (eds) *Tourism, Resilience and Sustainability: Adapting to Social, Political and Economic Change*, 103–115. Routledge, London.

Hooli, L. and Haaranen, A. (2019). What is the development in the "private sector in development" approach – Perspectives from Finnish enterprises. Submitted manuscript.

Hooli, L. and Jauhiainen, J. (2017). Development aid 2.0 – towards innovation-centric development co-operation: The case of Finland in southern Africa. In Cunningham, P. and Cunningham, M. (eds) *IST-Africa 2017 Conference Proceedings*, 1–9. IIMC International Information Management Corporation, Windhoek, Namibia.

Hooli, L., Jauhiainen, J., Järvi, A., Nkonoki, E., Taajamaa, V. and Käyhkö, N. (2019). Contextualising innovation in Africa: Knowledge modes and actors in local innovation development. In *IST-Africa Week Conference (IST-Africa) Proceedings*, Nairobi, Kenya, 2019.

Hooli, L., Jauhiainen, J. and Lähde, K. (2016). Living labs and knowledge creation in developing countries: Living labs as a tool for socio-economic resilience in Tanzania. *African Journal of Science, Technology, Innovation and Development* 8:1, 61–70.

IMF (International Monetary Foundation) (2019). *World Economic Outlook (April 2019)*. www.imf.org/external/datamapper/NGDP_RPCH@WEO/OEMDC/ADVEC/WEOWORLD. Retrieved June 2019.

ITU (International Telecommunications Union) (2019). *Data on the Internet Use and Mobile Phone Subscriptions in Africa*. www.itu.int/. Retrieved June 2019.

Jauhiainen, J. (2006). Multipolis – high technology network in northern Finland. *European Planning Studies* 14:10, 1407–1428.

Jauhiainen, J. (2017a). Asylum seekers and irregular migrants in Lampedusa, Italy, 2017. *Publications of the Department of Geography and Geology at the University of Turku* 7. University of Turku, Turku.

Jauhiainen, J. (2017b). Asylum seekers in Lesvos, Greece, 2016–2017. *Publications of the Department of Geography and Geology at the University of Turku* 6. University of Turku, Turku.

Jauhiainen, J. and Hooli, L. (2017). Indigenous knowledge and developing countries' innovation systems. The case of Namibia. *International Journal of Innovation Studies* 1:1, 89–106.

Jauhiainen, J. and Moilanen, H. (2012). Regional innovation system, high technology development and governance in the periphery: The case of Northern Finland. *Norwegian Geographical Journal/Nordisk Geografisk Tidskrift* 66, 119–132.

Jauhiainen, J. and Suorsa, K. (2008). Triple Helix in the periphery: The case of Multipolis in Northern Finland. *Cambridge Journal of Regions, Economy and Society* 1, 285–301.

Leke, A., Chironga, M. and Desvaux, G. (2018). *Africa's Business Revolution: How to Succeed in the World's Next Big Growth Market*. Harvard Business Press, Cambridge, MA.

Lemarchand, G. and Tash, A. (2015). *Mapping Research and Innovation in the Republic of Rwanda* (Vol. 4). UNESCO Publishing, Paris, France.

Levich, J. (2015). The gates foundation, ebola and global health imperialism. *American Journal of Economics and Sociology* 74:4, 704–742.

Mallory, A. and Giuliani, D. (2018). *Technology for Agriculture in Africa: A Fourth Revolution Without a Third?* https://briterbridges.com/agriculture-in-africa-a-fourth-revolution-without-a-third/. Retrieved May 2019.

Moon, E., Chang, K. and Min, B. (2016). An innovative approach to official development assistance to ICT for education in Bangladesh. *International Journal of Knowledge and Learning* 11:2–3, 178–189.

Murray, W. and Overton, J. (2016). Retroliberalism and the new aid regime of the 2010s. *Progress in Development Studies* 16:3, 244–260.

Nawrot, K., Juma, C. and Donald, J. (2018). African megacities as emerging innovation ecosystems. In Kleer, J. and Nawrot, K. (eds) *Rise of Megacities. The Challenges, Opportunities and Unique Characteristics*, 221–258. Barnes & Noble, New York.

Odusola, A., Cornia, G., Bhorat, H. and Conseicão, P. (eds) (2017). *Income Inequality Trends in sub-Saharan Africa. Divergence, Determinants and Consequences.* UNDP, New York.

OECD (Organization for Economic Cooperation and Development) (2019). *Net Official Development Assistance.* https://data.oecd.org/oda/net-oda.htm/. OECD, Paris. Retrieved June 2019.

Schillo, R. and Robinson, R. (2017). Inclusive innovation in developed countries: The who, what, why and how. *Technology Innovation Management Review* 7:7, 34–46.

Shizha, E. and Diallo, L. (2016). *Africa in the Age of Globalisation: Perceptions, Misperceptions and Realities.* Routledge, London.

Ratha, D., Mohapatra, S. and Plaza, S. (2008). Beyond aid: New sources and innovative mechanisms for financing development in Sub-Saharan Africa. *Policy Research Working Papers.* World Bank, Washington, DC.

SIDA (Swedish International Development Agency) (2015). *Support to Innovation and Innovation Systems – Within the Framework of Swedish Research Cooperation.* SIDA, Stockholm.

Tigabu, A., Berkhout, F. and van Beukering, P. (2015). The diffusion of a renewable energy technology and innovation system functioning: Comparing bio-digestion in Kenya and Rwanda. *Technological Forecasting and Social Change* 90, 331–345.

UNECA (United Nations Economic Commission for Africa) (2018a). *Africa Sustainable Development Report 2018. Towards a Transformed and Resilient Continent.* UNECA, Addis Ababa.

UNECA (United Nations Economic Commission for Africa) (2018b). *African Statistical Yearbook 2018.* UNECA, Addis Ababa.

United Nations (2015). *The 2030 Agenda for Sustainable Development.* United Nations, New York.

United Nations (2019a). *World Population Prospect 2019.* https://population.un.org/wpp/. Retrieved June 2019.

United Nations (2019b). *2018 Revision of World Urbanization Prospects.* United Nations, New York.

USAID (2019). *Catalyzing Innovation and Partnership.* www.usaid.gov/catalyzing-innovation-and-partnership/. Retrieved June 2019.

World Bank (2019). *Taking the Pulse of Africa's Economy.* April 8. World Bank, Washington, DC.

2

INNOVATION AND DEVELOPMENT IN AFRICA

Introduction

Innovation and knowledge are highly relevant to regional socio-economic prosperity worldwide. African countries are no exception, although in Africa, the factors that shape innovation and knowledge-creation capacities may be somewhat distinctive, and the geographies of innovation are very uneven. Innovation's positive contribution to economic progress is usually regarded as very straightforward. However, the relationship between innovation and both poverty and inequality – among the most fundamental development challenges in Africa – is much more complex. Novel context-specific and transformative policies are needed to address local development needs.

For a long time, innovations were understood to develop along linear and incremental pathways. A consensus prevailed that systematic R&D is necessary to generate innovations that ultimately increase general productivity. However, R&D investments are insufficient if the knowledge they produce is not appropriate to the market or it is not possible to translate R&D into novel and applicable products, services or organisational forms (McCann & Ortega-Argilés, 2015). Moreover, not all innovations emerge from systematic R&D. In fact, many innovations derive from doing, using and interacting (Jensen et al., 2007). Innovations emerge from multiple sources and forms, and may be the results of interaction between a variety of individuals, institutions and contexts. An interaction usually refers to the act of generating, sharing and applying knowledge among individuals and institutions (Doloreux, 2002). The networking and clustering of innovation-related actors may also lead into broader innovation ecosystems consisting of relationships supporting both research and commercialisation of innovations.

- In Africa, it is important to acknowledge different knowledge bases and development needs in the context of developing innovations.
- Innovation development is a social process that can be supported with adequate policies.
- To achieve more sustainable societies in Africa, the focus must be on inclusive innovations and transformative innovation policies.

In this chapter, we discuss the conceptual cornerstones of innovation development and their relevance to Africa. Section 2.2 begins with an overview of knowledge and innovations from a conceptual viewpoint. We explain how learning and practices transform knowledge into innovations. Section 2.3 explains innovation processes and how the Triple Helix and innovation system approaches can be used to develop innovations systematically.

In Section 2.4, we focus more specifically on how knowledge develops into innovations in the African context, and how innovation development is transformed into an innovation system in Africa. We also elaborate on the ways that the distinctive socio-economic contexts in Africa present challenges for innovations and their systemic development. In Section 2.5, we discuss the main challenges of innovations in development, and why the impacts of innovation-focussed development cooperation can sometimes be the opposite of the expected impacts. We highlight potentially transformative innovation policies, which are inclusive, participatory, and built around the needs and knowledge of local communities. Inclusive innovations help the poor and other marginalised segments of African societies, and promote more sustainable development. Inclusive innovations contain the implementation of novel ideas to create opportunities for economic and social well-being of the poor and help in tackling societal grand challenges (George et al., 2016). Effectively grounding innovation policies in local contexts and including local communities in the processes of innovation policymaking and implementation unlock major opportunities for innovation that are based on local indigenous knowledge and promote open and responsible innovation processes. Such approaches contribute to a transformative reorientation toward more sustainable societies in Africa. We conclude the chapter with Section 2.6, in which we present a better way of customising innovation policies to African socio-economic developments.

To innovation from knowledge via learning

Over the last several decades, knowledge has become a main asset for economic development. In the knowledge economy era, knowledge is the central source of innovation. Learning and novel interpretations are needed to create knowledge and transform knowledge into innovations.

Knowledge and its creation processes that aim at innovations have been a major topic of study, for example, in economics, geography, knowledge management, information and organisation science and innovation studies (Amin & Cohendet, 2004; Ibert, 2007; Carayannis & Campbell, 2009; Malecki, 2010; Hautala, 2011; Nonaka & Toyama, 2015). However, development policies, along with many studies regarding development and innovation, fail to properly define or conceptualise knowledge or innovation. Therefore, we present here an overview what knowledge is and how it develops through learning to innovations.

The ancient Greek philosopher Plato provided a classic definition of knowledge that remains useful. Plato defined knowledge as an individually interpreted justified true belief (Nonaka & Takeuchi, 1995: 21). This interpretation defines knowledge with reference to a previous understanding of something. This refers to the subjective human elements underlying knowledge creation and the transformation of knowledge into innovations. In creating new knowledge, it is common nowadays to have ICTs that facilitate person-to-person interaction. To draw out the new knowledge, these persons must understand each other, but must also have slightly different viewpoints (i.e. they must have some, but not too much, cognitive distance [Hautala & Jauhiainen, 2014]). The application of new knowledge requires an absorptive capacity based on existing knowledge, technologies and the human ability to recognise the value of knowledge and understand how to assimilate or apply its content (Cohen & Levinthal, 1990). The absorptive capacity of people and regions vary – for example, in accordance with human competence and existing knowledge, which depends, for example, on backgrounds in education and experiences.

The justification possibilities of knowledge vary, as knowledge has dimensions that are more or less explicit and implicit. Explicit knowledge is communicable from person to person via a systematic set of codes, such as through a model, language or pattern (Nonaka & Takeuchi, 1995: 59). There is little room for interpretation, facilitating discussions about knowledge transfer. Therefore, explicit knowledge can be shared across distances and beyond territorial and socio-cultural borders. Science, technology and innovation are tools to achieve new knowledge. They are also outputs of such knowledge creation processes (Lundvall, 2004; Cozzens & Kaplinsky, 2009; Carayannis & Campbell, 2009; Lundvall, 2016).

At the other end of knowledge conceptualisation is the tacit knowledge based on human understanding, interpretation and practices. According to Hungarian philosopher Michael Polanyi (1881–1967), people know more than they are able to tell. Many individuals neither need nor are able to conceptualise something they know and are able to do, nor code that into explicit knowledge. Tacit knowledge combines personal interpretation, individual working practices and interaction with others. Tacit knowledge requires close proximity, as it is usually context specific (Polanyi, 1966). It is therefore difficult to transfer from person to person, and impossible to transfer from one region to another. In practice, knowledge creation consists of a combination of the explicit and tacit dimensions of knowledge. Both modes are necessary in a knowledge-based economy and society.

Another way to approach knowledge is to consider to how knowledge is created and where it is meant to be used. From this perspective, knowledge can be divided into analytical, synthetic and symbolic knowledge (Asheim & Gertler, 2005; Asheim & Coenen, 2005). Analytical knowledge refers to knowledge generated through existing scientific knowledge and transformed from systematic R&D into science. It is abstract, somewhat universal and usually codified into scientific laws or formulas. This knowledge creation is mainly based on formal models, codified science and rational processes. Such codified knowledge can be mediated through different ICTs, as mentioned previously regarding the notion of explicit knowledge. For analytical knowledge, the geographic proximity of actors is less relevant than it is for other knowledge bases. The application and diffusion of this knowledge relies on the absorptive capacity of the receiving country, its institutions and skilled individuals. Analytical knowledge plays an important role in high technology and companies operating in advanced technological fields, such as artificial intelligence, nanotechnology or biotechnology.

Synthetic knowledge refers to new combinations of societally relevant existing knowledge and their use. The intention is to use this knowledge to solve concrete practical problems in society. Such knowledge is found, for example, in engineering, machinery and industries. The investment in R&D is less important than incremental product and process development, which is accomplished by testing, experimentation and practical pursuit of work and applied research. This amplifies the tacit dimension of synthetic knowledge (Asheim & Gertler, 2005).

Symbolic knowledge refers to the aesthetic dimension of products, the development of images and designs, and the economic use of cultural objects. It is based on a context-specific creation emerging from craft, practical skills and searching skills, which maintain a strong tacit component (Asheim et al.,

2007). In the 21st century, symbolic knowledge has become a very important dimension in making a difference among customers who may potentiality select from among a very similar set of consumption goods, whether clothing or high technology. Symbolic knowledge is often needed to aesthetically "cover" the products created through analytical and/or synthetic knowledge, thereby assigning distinctive aesthetic value to the product.

Knowledge and innovations are intertwined. Though innovations have always been part of human development, it was not until the early 20th century that their conceptual discussion became so important. Austrian economist Joseph Schumpeter (1883–1950) defined innovation as a new or essentially improved product, process, market, raw material or organisational arrangement (Schumpeter, 1934). Like with Plato's conception of knowledge as a novel justified interpretation, Schumpeter's idea of innovation also stresses the importance of novelty. Whereas in Plato's definition, novelty is tested by the academic community, in Schumpeter's definition, novelty is tested by the market. Innovations address potential users' needs with technological opportunities. New definitions of innovation have emerged during the late 20th and early 21st centuries (see Box 2.1).

BOX 2.1   INNOVATION

Innovation is a new or essentially improved product, process, market, raw material, or organisational arrangement. New inventions emerge each day, but an invention becomes an innovation only if it is implemented and applied successfully in the market and/or society.

Innovations are not always based on original inventions. They may be successful modifications of existing products or services, such as an application of an old innovation in a new context. In such an instance, existing knowledge is interpreted and applied differently. Incremental innovations take the existing products and services as their base, slightly modifying and improving them. However, radical innovations represent substantial departures from existing product or services. They may fundamentally change the whole context and release completely different products and services or modes of providing them. The impacts of innovation vary.

From the late 20th century onward, many new innovation-related concepts have emerged to better articulate the character and process of innovation in the global knowledge-based era (see Pansera & Martinez, 2017). Such concepts include process innovation, design innovation, platform innovation, service innovation, inclusive innovation (see Chataway et al., 2014), open innovation (see Von Hippel, 1990), frugal innovation (see Section 2.6 and Chapter 8), need-based innovation (see Chapter 5) and responsible innovation (see Chapter 8).

A useful and widely used mode to explain how the creation of knowledge leads to innovations involves distinguishing between the STI mode (science, technology and innovations) and the DUI mode (doing, using and interacting) of learning (Jensen et al., 2007; Isaksen & Karlsen, 2010; Binz & Truffer, 2017; Blažek & Kadlec, 2018). This division is not so clear in real everyday practices. Usually, innovations are composed of combinations of different knowledge bases with diverse emphases. The combination of different learning modes has a stronger influence on innovation outputs than either of the individual modes in isolation (Parrilli & Heras, 2016).

The STI mode of learning relies on formal learning processes and R&D activities. It requires investments in talented human resources in science, technology and related infrastructure. Explicit knowledge is fundamental for the STI mode of learning, thereby making it appealing for innovation policies and technology transfer. This learning mode utilises mainly analytical knowledge and is present in both incremental and radical innovations. The STI mode of learning requires advanced education systems and solid technological infrastructure.

The DUI mode of learning is often applied through a problem-solving process. This learning mode involves context-specific tacit knowledge, as it is based on interaction among knowledge producers, appliers and users within a geographical proximity. This mode of learning is typically based on synthetic and symbolic knowledge. In enterprises, innovations based on the DUI mode of learning depend not only on internal processes, also on the capacity of enterprises to interact with users, suppliers and competitors (Asheim et al., 2016). In contrast to the STI mode of learning, context-specific innovations are more difficult to transfer or scale to another context.

Innovation processes and innovation system frameworks

Globally, the geography of innovations is very uneven (Shearmur et al., 2016). Innovations emerge in specific places, and there are many areas with very limited innovation outputs. Population size and amount of natural and economic resources do not automatically translate into the emergence of innovations. The socio-economic context imposes both push and pull factors on innovation development. Innovations tend to concentrate in certain areas, mostly in the Global North. The amount and quality of skilled individuals, technological and knowledge infrastructure, and well-functioning basic infrastructure are important. Moreover, as innovations often derive from highly interactive processes and require tacit knowledge, the physical proximity engendered by co-location of the key actors and resources enables knowledge sharing and creation processes (Hautala & Jauhiainen, 2014). Therefore, while innovation can be defined in similar conceptual terms everywhere, the processes of innovation are highly distinctive across different contexts, and especially between the economic core and periphery (Hautala & Jauhiainen, 2019).

Over the past decades, the purposeful development of innovations has become a key activity in the public and private sectors, as well as among non-governmental organisations, various communities and individuals in the Global North and, increasingly, in the Global South. Innovations are developed in networks and interactions between different actors (Asheim et al., 2016). In such processes, innovations are thought to develop through more complex, non-linear, evolutionary, and cumulative knowledge creation and learning processes between various actors (Pavitt, 2002).

The key comprehensive framework for developing innovations is the innovation system approach (see Box 2.2), which emerged in the 1980s and has continued to develop since that time (Freeman, 1987; Lundvall, 1992; Nelson, 1993). An innovation system consists of the assemblage of all significant economic, social, political, organisational, institutional and any other factors affecting the development, use and diffusion of innovations. There are also different geographical extensions of an innovation system, from local to regional, national and international. There is a distinctive character of the innovation process across nations, regions and sectors, due to diverse technological and institutional trajectories (Lundvall, 1992). The definition of an innovation system depends on whether it is considered to be a narrow or a broad system (Asheim et al., 2016).

The principles for a narrow innovation system can be explained through the Triple Helix model that has become a very popular approach to discussing the possibilities and practices involved with the systematic development of innovations. The Triple Helix refers to an increasingly intertwined purposeful co-operation between the private sector, universities and the government (public sector) to gain innovations through novel production, and the transfer and application of knowledge (Etzkowitz & Leydesdorff, 1995, 2000). The Triple Helix studies and maps the models of cooperation occurring between various institutions and related processes of governance in many countries. Researchers identified several relationship types through analysis of Triple Helix innovation activities: technology transfer, collaboration and conflict moderation, collaborative leadership, substitution and networking (Ranga et al., 2013). These align the interests of these institutions and provide ways to design and reform a more efficient and comprehensive model of how to achieve innovations in a knowledge-based society.

The private sector was seen as the main actor in development, R&D investments and commercialisation of innovations (see Chapter 5). The government should support innovations through incentives, regulations and laws governing innovation-related activities and proper innovation policies. Universities are not only intended to generate novel basic and applied knowledge, but, through their students, they are an endless source of new human capital. To enhance the Triple Helix, universities should become more entrepreneurial, foster new spin-off companies and focus on innovations originating from applied research (Schot & Steinmueller, 2018; see Chapter 6).

Initially, the Triple Helix was studied and discussed in the contexts of the more developed countries. Eventually, the discussion also integrated some African countries (Etzkowitz & Dzisah, 2007; Mêgnigbêto, 2013; Patra & Muchie, 2018). The Triple Helix also came under fire for its failure to consider many relevant actors, such as civil society and the informal sector. In the early 21st century, an updated model of the Triple Helix emerged; this was the Quadruple Helix, which was intended to cover the aforementioned negligence by incorporating media-based and culture-based public sectors and non-governmental organisations into the model (Carayannis & Campbell, 2009).

Broadly understood, an innovation system involves cooperation among all economic, social, political, organisational, institutional and other factors affecting development, application and diffusion of innovations (Lundvall, 1992; Edquist, 2005; for the African context, see Figure 2.1). The broad understanding of the innovation system has been criticised for its fuzziness. In this loose network, it is difficult to analyse the precise interactions between different actors of an innovation system and the impacts that these interactions have (Markusen, 2003). However, as we explain in the following sections, this broad understanding of the innovation system is especially relevant to Africa, where most innovations do not derive from universities and their science-based research.

BOX 2.2 INNOVATION SYSTEM

An innovation system (see Figure 2.1) is the assemblage of all signifi-cant economic, social, political, organisational, institutional and any other factors affecting the development, use and diffusion of innova-tions (Edquist, 2005). The concentration of innovation-related actors supports the formal and informal communication and learning among these key actors. It consists of innovation system components (for exam-ple, actors, as mentioned previously), linkages (themes of innovation system activities) and boundaries (connections within and outside of the innovation system) (Coenen et al., 2017). In the innovation system, innovation is a complex, non-linear and interactive process that exists in multiple forms. Innovation emerges from interaction and co-operation between various types of actors in the innovation system. Some actors deal with generation of new knowledge and its diffusions. Other actors then apply and exploit this novel created knowledge. The role of inno-vation policies is to support and regulate the innovation process.

Innovation systems usually have a geographical scope in their organi-sation and governance, as well as in knowledge creation, application and diffusion. Innovation systems have been labelled "national" (Lundvall, 1992), "regional" (Cooke, 1992; Asheim, 1996; Braczyk et al., 1998), "macro-regional" (Bellini & Hilpert, 2013) or "cross-border" (Trippl, 2010; Lundquist & Trippl, 2013), to emphasise their specific geographi-cal features. Moreover, local innovation dynamics have been analysed, for example, in the context of innovation milieus (Moulaert & Sekia, 2003) and clusters of innovation.

A well-functioning innovation system is supported by a critical mass of human capital in a well-structured organisational and institutional setting. Therefore, most successful innovations emerge from specific geographical areas, predominantly in areas in the Global North, such as Silicon Valley in the United States and the Stuttgart region in Germany, but also in places like Singapore and the Jiangsu province in China; this makes global innovation geography very uneven. However, innovations can also emerge in peripheral areas (Jauhiainen & Suorsa, 2008), in the Global South (Ndabeni et al., 2016; Hooli et al., 2019) and in cases of temporary proximity to innovation-related actors (Bathelt, 2011; Carrincazeaux & Coris, 2011), facilitated by the modern ICTs.

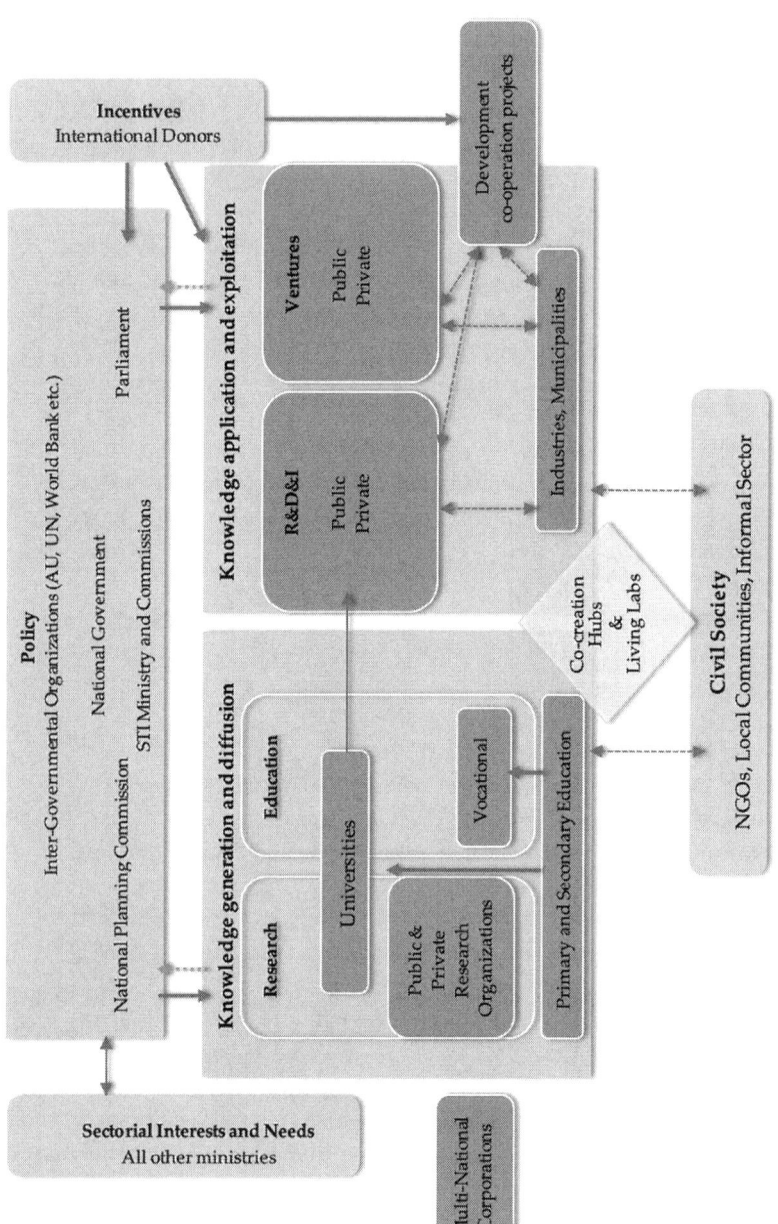

FIGURE 2.1 Innovation system in the African context

Innovation development is context-specific and embedded into a certain territory. However, regions and countries with successful innovation systems share similar socio-economic features. Among these are, for example, good and effective education systems, vigorous foundations of knowledge creation in technology, science and business, high-level commercialisation of knowledge and remarkable venture capital investments and flexible labour markets (Tödtling et al., 2013). The recent innovation literature underlines the importance of openness to the innovation system's components, linkages and boundaries. Knowledge creation must be interactive and continuous, with human resource formation and capability building, the fostering of combined related competencies with new combinations of knowledge, and the provision of contact points for extra-regional and extra-national networks (Coenen et al., 2017). This allows for the co-creation of innovations with many actors, including innovation users and end-customers (Garcia Haro et al., 2014). However, to transfer innovation systems into innovation policy is a complex process (Frenken, 2017).

In the 2010s, the concepts and policies for smart specialisation as an extension of innovation systems, and also broader discussion about innovation ecosystems, emerged. Smart specialisation refers to the capacity of an economic system – for example, a locality, region or even a nation – to generate new specialties of innovative economic development. These are found through the discovery of new domains of opportunity, creating competitive advantage, and the local concentration and agglomeration of resources and competences in these opportunity domains (Foray, 2014). In broader innovation ecosystems, the emphasis was also on considering commercialisation of innovations. It was seen that the focus of traditional innovations systems is more on public-sector-driven knowledge and innovation generation, rather than the market-driven exploitation and commercialisation of innovations. The efficiency of an innovation ecosystem was often measured simply by calculating the investments into R&D, and the economic value and profits resulting from the innovation induced changes deriving from the goods and services enhanced with the R&D. The geographical nearness supported the development of innovation ecosystems, so the national and local public policies and planning were important as well.

Innovation systems and development in the African context

In the context of the expanding knowledge-based economy in Africa, it has become increasingly important to approach innovations systematically. Such an approach has a strategic value for industrial and development policies (Bartels et al., 2012) and in attempts to catch up the Global North (Arocena & Sutz, 2000). A key element in the systematic development of innovations is a national innovation system to accelerate economic performance and increase competitiveness. In general, an innovation system in Africa comprises, in principle, the same elements and key actors as in the Global North (Figure 2.1).

However, most African countries, regions and institutions still follow the resource-dependent and commodity-driven development pathways. In their development strategies, many African countries aim to move from the current economic dominance of agriculture to industry. Nevertheless, it has been argued that several African countries could skip industrialisation and leapfrog directly to the knowledge economy (Oyelaran-Oyeyinka, 2014). This would require a move away from the established development path of the former industrial countries of the Global North that made this shift in the 19th and 20th centuries. Innovations are not developed in a vacuum; past development trajectories influence the current opportunities to develop innovations (Boschma, 2015).

As mentioned previously, an innovation system is a systematic approach to develop innovations. It is conceptually universal, but its implementation varies across different countries. Successful innovation systems or related innovation policies cannot be directly imitated or transferred (Tödtling & Trippl, 2005). Therefore, an innovation system in an African country must be organised differently than in more economically and technologically advantaged countries (Lundvall et al., 2009). African countries must set targets and priorities in substantially different ways than the wealthiest countries.

The need for a distinctive mode of innovation system formation is not a negative issue, because a country and region's unique assets and characteristics may create an advantage. From these, it is possible to build competitive advantages through smart specialisation that derives from place-based policies that enhance a given region's economic variety and sharpen its competitive advantage (McCann & Ortega-Argilés, 2013). Further, it has more recently been claimed that innovation systems should be open connected to the global value chain (Coenen et al., 2017; Jurowetzki et al., 2018). There are always inflows of human and financial capital and knowledge from external sources. This is especially relevant for emerging or incomplete innovation systems in Africa, as external sources, such as multinational companies or development cooperation projects, can replace some missing or dysfunctional parts of an innovation system.

A major challenge in the formation of innovation systems in the African countries is the organisational and institutional thinness of the innovation system in Africa (Hooli & Jauhiainen, 2017). Organisational thinness is the absence or insufficient quality and amount of skilled individuals or critical mass in companies, R&D laboratories, universities, government, associations, unions or any other organisations relevant for innovation processes (Moodysson & Zukauskaite, 2014). Regarding knowledge based on the STI mode of learning, African universities have limited resources to conduct high-level international research (see Chapter 6). They are seldom included in governments' development strategies, and their resources focus largely on education at the graduate level (Bartels et al., 2016). From an operational perspective, the key networks among various actors in innovation processes in Africa are weak. Hence, the knowledge exchange between actors from different sectors is very limited, which hinders the interaction and mutual learning required for innovation processes (Pavitt, 2005). In addition, a major obstacle in Africa is the general lack of professionals and individuals acquainted with the systematic and comprehensive nature of innovation and innovation systems (Adebowale et al., 2014).

Another challenge is institutional thinness, which refers to the insufficient quantity, quality or absence of formal institutions (rules, regulations and laws) and informal institutions (values, norms, and other assets important for cooperation and innovation) that stimulate mutual learning and knowledge exchange (Moodysson & Zukauskaite, 2014). It is very common for African innovation development to suffer from a lack of cooperation, insufficient funding and poorly managed operations. In addition, related policy, institutional, legal and regulatory environments have been unformed for too long in many contexts, at least in the past (Oyelaran-Oyeyinka, 2006). Systematic innovation development is rarely included in the national development strategies with any explicit plan for implementation; when such developments are included, they are seldom contextualised to local socio-economic contexts (Watkins et al., 2015). The regulatory quality and efficiency of the national governments are distinctive institutional features that have major impacts on innovation development in the African regions (Oluwatobi et al., 2015).

Moreover, the macro-regional cooperation patterns in Africa are still scattered, and the small and growing local economies have been unable to attain adequate financial, technological or human capital needed for a well-functioning innovation system (Hartzenberg, 2011). Therefore, innovation policies in Africa must be aimed at macro-regional integration and economic cooperation (Scerri, 2013). In recent years, there have been efforts to open intra-African trade with free trade areas, for example. This may also be positive for the development of innovations and the creation of a more systematic innovation system. A discouraging factor is identified by Barasa (2018), who found a positive relationship between corruption and innovation in Africa. The specificity of physical assets increases the likelihood of innovation in a business environment characterised by a high degree of corruption, which is to say that corruption apparently "greases the wheels" of innovation processes in Africa (Barasa, 2018).

Another difference between the innovation systems and more general innovation developments between the Global North and Africa is the emphasis on different modes of learning that lead to innovations. In Africa, the STI mode of learning can be fundamental in the deepening era of digitalisation, as with, for instance, new technologies in medicine, logistics, agriculture and digitalised services. Despite the incremental nature of innovations that derive from the STI mode of learning, there are some notable exceptions that foster an especially broad transformative change. For example, Aerobotics is a South African drone software company that develops a data-analysis platform, based on artificial intelligence, to help farmers optimise their activities. Another example is the Ethiopian Safe Delivery app, which has been developed to coach women to give birth, as the vast majority of births there occur outside of institutional health facilities. A further example, originating from the STI mode of learning, is the biomedical application MamaOpe in Uganda. It was developed by two former Makerere University students, together with international partners. MamaOpe uses a smart jacket to diagnose and monitor pneumonia patients; it identifies pneumonia symptoms approximately four times faster than doctors, while also eliminating human errors. It is an especially relevant innovation in Africa, where, according to UNICEF estimations, around 500,000 children die annually from pneumonia and there is a severe shortage of doctors. Another transformative innovation has been the Kenyan mobile money transfer system M-Pesa, which has opened financial services up to millions of people, elevated thousands out of poverty, and substantially increased financial transparency (see Chapter 5 and Box 5.4).

Despite these promising examples, in general, innovations based on systematic R&D and analytical knowledge in Africa are scarce, especially relative to the size of the continent's population. Additionally, many most advantageous innovations are owned or developed by foreign companies or multinational corporations. For example, among other companies from the United States, Black Star Energy and Zola Electrics build up solar-powered infrastructure and transport equipment to rural areas in Ghana, Nigeria, Rwanda, Sierra Leone and Tanzania to electrify villages without either connection to energy grids or financial institutions. Part of the innovation-oriented activities take place in cooperation between local and foreign experts. Also, like everywhere else in the world, most private-sector innovations are consumer goods that target middle- and high-income consumers but do not have any goals to alleviate broader social concerns.

Nevertheless, the situation may change in the near future, as a new generation of home-grown innovation developers emerges in Africa and the global start-up culture flourishes (see Chapter 5). The home-grown start-ups are particularly important, as they are much closer to local communities, as well as their needs and challenges (see Chapter 8), relative to international organisations based in Africa (see Chapter 7). In general, Africa needs more innovations based on analytical STI mode of knowledge (see African Union, 2015). However, focussing national innovation policies and innovation systems more toward STI-based innovations that lack competitive advantages and economic and human capital may lead to expensive policy mismatches and developmental lock-ins.

Despite the attraction of science and technology as the main bases of innovation systems, in the African context, innovations generated from applied and symbolic knowledge and through the DUI mode of learning are especially relevant (Kraemer-Mbula & Wamae, 2010). The high diversity of the contexts, cultures, societal organisations and languages in Africa makes the DUI mode of learning especially full of potential. This commands acknowledgement of local, indigenous and traditional knowledge. It has been a rather unused innovation opportunity in Africa. It is especially important because it is often the main asset of the global poor (e.g. Gorjestani, 2004). Indigenous knowledge features in some national innovation policies and the construction of innovation systems in Africa (Jauhiainen & Hooli, 2017). For example, the Multidisciplinary Research Centre of the University of Namibia runs several research programmes aiming to create added value from indigenous knowledge (Chinsembu, 2015). Some projects are based on analytical knowledge and R&D, like endeavours to exploit indigenous plants to develop compounds for anti-malaria medicine and treatment for HIV-related conditions and various food and agricultural products. On the other end, symbolic knowledge has been especially important in the African film and music industries, for example in Nigeria, as well as in tourism in many African countries. The role of indigenous knowledge in African innovation system development is discussed, in more detail, in Chapter 8.

From innovation challenges towards transformative innovation policies in Africa

The general attitude towards innovations in Africa is positive; they are seen as an opportunity for future economic growth. However, there are threads that can result from innovations in Africa but are rarely addressed. This is partly because only a few researchers have studied the broader implications of innovations for local societies in Africa and the people whom innovations specifically target (Lee, 2016). While innovations definitely present positive opportunities for the continent's social and economic transformation, they also bring along challenges. Researchers have become increasingly concerned about the negative externalities of the existing innovation models in the Global South (Cozzens & Kaplinsky, 2009; Soete, 2013). To turn away from possible negative outcomes, novel and responsible transformative innovation policies and inclusive innovations are needed (see Chapter 8).

Schot and Steinmueller (2018) note that the dominant approaches to innovation in the Global North are insufficient to address the challenges of inequalities, poverty and the environment that characterise Global South. Kaplinsky (2011) has even argued that innovations contribute directly to the expansion of inequalities (see also Schillo & Robinson, 2017). On the one hand, innovations benefit the better-off population in the Global South, thereby increasing their wealth and making some individuals super-rich. In addition, innovations may be linked to the expansion of geographical inequality and segregation. Wealthy skilled individuals in Africa are mostly agglomerated within countries' capitals and other economically relevant urban areas, and they live in affluent neighbourhoods there. Large cities have become better places to create and consume contemporary innovations. The changing patterns of global value chain dynamics, the tendency of innovation and production activities to cluster in developing economies and accelerate uncontrolled urbanisation there, and the bias favouring urban over rural areas has widened the inequality between regions in Africa.

In fact, there is a strong agglomeration of innovations in a few particular African regions and cities. In Africa, most innovation activities are agglomerated in a few regions, such as the Western Cape (Cape Town) province of South Africa, the Cairo metropolitan region in Egypt, the Lagos metropolitan region in Nigeria and the Nairobi metropolitan region in Kenya. These are also areas in which analytical knowledge is created in the top-level African universities (see Chapter 6, Table 6.1, and Figure 1.4). However, knowledge creation, especially with tacit knowledge, is sticky and embedded within its socio-economic context, and it is difficult to diffuse and transfer it from these contexts to elsewhere (Tödtling & Trippl, 2005).

On the other hand, innovations replace existing technology and cut out low- and mid-skilled work, at a significant cost to incumbent workers and companies (Bogliacino & Pianta, 2010). Innovations do not usually reduce employment opportunities for population segments with the most advanced skills, who can more easily apply their competencies in changing circumstances and actually benefit from innovations. Nor will innovations so crucially impact the employment of the low-skilled population because their participation – as cleaners, waiters and other basic service providers – can rely on demand in the future. In fact, innovations most commonly reduce the need for medium-skilled employees with jobs that can be automated due to innovation processes and the viability of replacement with inexpensive new technologies (Lee, 2016). Such creative destruction hits hardest in those countries of the Global South where old technologies and processes demand a large workforce at the bottom-end of the global value chain. Global corporations have moved their employment-intensive production operations to these countries, and eagerly reduce their fixed costs if innovation provides an opportunity for cost-cutting.

An additional challenge is that the costs of innovation development are often mostly covered as public expenditures. Although public sector in most African countries invests very little in R&D and other innovation-supporting activities, the private sector invests even more scarcely. The AU has set a target, encouraging African countries to invest 1% of their GDP in R&D. Thus far, none of the African countries have reached this target. South Africa's R&D investments have come closest (0.8% of the GDP) to the target, and its absolute investment amount is overwhelmingly the largest in Africa. Investments in innovation-supporting activities, like R&D facilities and technology centres, require substantial financial resources, large-scale infrastructure, skilled labour and availability of inexpensive energy and natural resources. It is not clear whether increased investment in an innovation system and R&D can generate faster economic and social catch-up in the Global South (Schot & Steinmueller, 2018). In fact, increasing investments in R&D and innovation activities in African countries may leave fewer public resources available to meet the immediate needs of the poor. Therefore, support for such innovations may even increase inequality and poverty; at least, these may be short-term impacts in the Global South. Highly innovative places in the Global North are often characterised by high levels of inequality. Innovations have not reduced inequality, even if innovation systems function well there.

While technological innovations sometimes improve the everyday life of people, they do not necessarily result in a more sustainable African society. Innovations also contribute to resource-exhaustive and unsustainable mass production. Such production is often still based on wasteful consumption of fossil fuels and other natural resources (Steffen et al., 2015). Innovations are also the root of many broader global challenges, like wars fought via weapon innovations, climate change exacerbated by innovation of internal-combustion engines, or environmental and health concerns through harmful asbestos produced by a housing insulation innovation.

In Africa, many mass-produced items are targeted mainly at consumers with extensive purchasing capabilities. Due to a lack of resources, frugal innovations (see Chapter 5), like small upgrades or improvements with the downgrading of existing products, certain services or processes, have become popular by reaching larger segments of the population – one out of three Africans live below the poverty line. Frugal innovations are often mass-consumption products that can be offered at substantially lower prices. The reduction of price results in stripped technological features and reduced product quality, for example, in cars and mobile phones. Such changes in the products may also create challenges for development.

The concerns about innovations' relation to development are especially relevant in the context of international development cooperation, which increasingly focuses on creating innovations as part of the development aid. The relationship between innovation that is often geared around economic growth and the general objectives of development cooperation, such as alleviation of poverty and inequality, is complex (Altenburg, 2009).

Innovation development requires long-term targets, consistent policies, economic and social capital and the ability to take and sustain risks. This is challenging for many aid-receiving countries and their communities. In the short term, innovation-focussed development cooperation may actually mostly benefit the wealthy and highly educated, who are most capable of taking part in these innovation-related processes. This is a small segment of population in these countries. In 2014, it was estimated that, in all of Africa, there were about 169,000 millionaires, and their number was expected to grow to 260,000 by the mid-2020s. Despite this relatively rapid growth compared with the United States, the larger number in Africa still represents around 2% of current millionaires in the United States. In addition, in South Africa, the wealthiest country of Africa, the number of millionaires has declined in the latter part of the 2010s (Knight Frank, 2019).

Nevertheless, innovations have become a common feature in African countries' development strategies and policies. Most development strategies in Africa take innovation and development policies for granted. Such strategies are legitimised by universal economic rationales that expect improved employment and novel technology to resolve the main challenges in African countries and new technologies to be applied in socially and environmentally sustainable ways. Thus, there is a need to contextualise innovation activities to local conditions and sustainable development needs and to implement better place-specific policies.

In fact, the wider public and an increasing number of researchers have, over the last years, become engaged in broader discussions about technological options and directions for innovation policies that may better support socially and environmentally sustainable development in Africa (Muchie et al., 2003; Lundvall et al., 2009; Kaplinsky, 2011; Schot & Steinmueller, 2018; Diercks et al., 2019). Moreover, it has started to become evident to the international development community that a broader socio-economic transformation is needed in Africa to achieve a genuine sustainable development, rather than relying on the outcomes gained by the technological improvement of industry (Chataway et al., 2017). The aspiration for such global outcomes is also evident in the 17 SDGs formulated by the United Nations (2015).

Schot and Steinmueller (2018: 1562) argue that new innovation policies should not be designed to produce more consumption goods. Instead, transformative innovation policies should prioritise the systems-wide transformation toward social, environmental and economic sustainability and inclusive societies (see Chapter 9). In this context, transformation refers to a broader and deeper change in the complex socio-environmental system instead of radical or incremental technological solutions. For example, innovation policies should not target technological change that would result in longer-lasting batteries for electric cars, if the mobility system is still dominated by unsustainable private motoring. Instead, the focus must be on reducing the need for transportation and supporting public transportation, bicycling and walking. The co-productive technological, social, and behavioural change should take place in an interrelated manner. Ultimately, a profound change is required in industry structures, infrastructures, regulations, products, cultural predictions, skills and user preferences.

According to Cozzen and Kaplinsky (2009), innovations may positively impact local community development, inequality and poverty, if innovation policies are strategically designed to address such objectives and if they are, accordingly, operationally implemented. Innovation development must include bottom-up approaches and innovation policies should support locally created innovations. Dutrénit and Sutz (2014) and von Hippel (2015) emphasise the need for participatory processes to democratise knowledge and innovation. Innovation processes should be open to all stakeholders, including members of even the most marginalised populations.

These novel innovations require a socially and culturally inclusive transformative innovation system that produces inclusive innovations (Chataway et al., 2014; see Chapter 8). Inclusive innovations originate from the creative power and problem-solving capacity of the poor. These may be products, services, business models, institutions, processes and supply chains aimed at overcoming poverty and new to the context (George et al., 2012). They are based on the needs of disenfranchised members of society, and they strengthen their economic and social conditions (Heeks et al., 2014). As a systemic perspective, inclusive innovation combines redistribution to the poor with active participation by the poor. It is a trajectory in which innovation activities benefit the poor by empowering them to become active stakeholders of the innovation system and in the pro-poor growth. It applies participatory processes to face the challenges to resolve conflicts with related strategies and practices (Johnson & Andersen, 2012).

Relevant and societally important innovations can be generated through the novel contextualisation of products into in Africa. Modified imitation or partial technology transfer can be economically significant behind the technological frontier. The key concern is not necessarily to provide a universal solution, but to offer an incremental improvement that fits into the local context. They can be, for example, new techniques in agriculture that yield an improved harvest, or the emergence of day care in regions where childcare has previously tied women to the home. Many such innovations remain invisible, in the sense that they lack a recognisable brand or web sites.

So far, the innovation system approach has not adequately focussed on the special challenges of the Global South. However, the potential for innovations to help eradicate poverty remains. In 2019, for the first time in the 2010s, the amount of poor people started to diminish in Africa (Hamel et al., 2019). An increasing number of good local innovative practices already exist in Africa and are discussed more specifically in Chapter 8.

Conclusions

Innovations are present everywhere in Africa. Innovations and knowledge are highly relevant to socio-economic prosperity in Africa, but geographies of innovation and factors shaping innovation and knowledge creation capacities are very uneven. Some innovations become viral applications that change the lives of millions of African people, while others remain local-level adjustments that help the everyday life of the poor. Africa is facing unprecedented and rapid social and environmental changes now and in the coming decades. African countries and regions need better innovation policies and more sustainable implementation of these policies to effect a broader transformative change.

Innovations must be developed systematically, and an innovation system is a tool to achieve this. Many African countries have launched an innovation system as part of their national innovation policies. However, in many cases, the main structure and contents of a given innovation system have been loaned, if not copied, from wealthier and economically and technologically more advanced countries in the Global North.

In Africa, the DUI mode of learning that aims at innovation can create a competitive advantage for African countries and communities. It is a meaningful tool to achieve inclusive societal development. However, the STI mode of knowledge is also needed to create cooperation and external links with key innovation actors in the Global North. There is potential to create agile innovation systems, which feature DUI modes and STI modes of learning. Ultimately, proper innovation policies and innovation systems are must be inclusive, participatory and societally transformative to eradicate poverty and bring social and economic prosperity to Africa. Grounding innovation policies in local contexts and engaging local communities in transformative innovation processes unlocks opportunities for innovation that are based on local and indigenous knowledge, while supporting open, responsible innovation processes. This approach lends itself to the opportunity to create more sustainable societies in Africa and beyond, in the Global South (see Chapter 9).

- An innovation may be similarly defined conceptually everywhere, but the processes of innovation are significantly distinctive across different contexts.
- Innovation development is a social project that can be supported and steered with adequate policies.
- In Africa, the focus must be on pro-poor inclusive innovations and transformative innovation policies.

Discussion questions

- What is an innovation and innovation system?
- What opportunities and challenges exist in the creation and implementation of a national innovation system in Africa?
- Find out a recent innovation originating from Africa and discuss its impacts to the rest of the world.

References

Adebowale, B., Diyamett, B., Lema, R. and Oyelaran-Oyeyinka, O. (2014). Introduction. *African Journal of Science, Technology, Innovation and Development* 6:5, v–xi.

African Union (2015). *Agenda 2063: The Africa We Want*. www.au.int/. Retrieved June 2019.

Altenburg, T. (2009). Building inclusive innovation systems in developing countries: Challenges for IS research. In Lundvall, B., Joseph, K., Chaminade, C. and Vang, J. (eds) *Handbook of Innovation and Developing Countries: Building Domestic Capabilities in a Global Setting*, 33–56. Edward Elgar, Cheltenham.

Amin, A. and Cohendet, P. (2004). *Architectures of Knowledge: Firms, Capabilities, and Communities*. Oxford University Press, Oxford.

Arocena, R. and Sutz, J. (2000). Looking at national systems of innovation from the South. *Industry & Innovation* 7:1, 55–75.

Asheim, B. (1996). Industrial districts as 'learning regions': A condition for prosperity. *European Planning Studies* 4:4, 379–400.

Asheim, B. and Coenen, L. (2005). Knowledge bases and regional innovation systems: Comparing Nordic clusters. *Research Policy* 34:8, 1173–1190.

Asheim, B., Coenen, L. and Vang, J. (2007). Face-to-face, buzz, and knowledge bases: Sociospatial implications for learning, innovation, and innovation policy. *Environment and Planning C: Government and Policy* 25:5, 655–670.

Asheim, B. and Gertler, M. (2005). The geography of innovation: Regional innovation systems. In Fagerberg, J., Mowery, C. and Nelson, R. (eds) *The Oxford Handbook of Innovation*, 291–317. Oxford University Press, Oxford.

Asheim, B., Grillitsch, M. and Trippl, M. (2016). Regional innovation systems: Past – present – future. In Sheamur, R., Carrincazeaux, C. and Doloreux, D. (eds) *Handbook on the Geographies of Innovation*, 45–62. Edward Elgar, Cheltenham.

Barasa, L. (2018). Corruption, transaction costs, and innovation in Africa. *African Journal of Science, Technology, Innovation and Development* 10:7, 811–821.

Bartels, F., Koria, R. and Andriano, L. (2016). Effectiveness and efficiency of national systems of innovation: A comparative analysis of Ghana and Kenya. *African Journal of Science, Technology, Innovation and Development* 8:4, 343–356.

Bartels, F., Voss, H., Bachtrog, C. and Lederer, S. (2012). Determinants of national innovation systems: Policy implications for developing countries. *Innovation Management Policy and Practice* 14:1, 2–18.

Bathelt, H. (2011). Innovation, learning and knowledge creation in co-localised and distant contexts. In Pike, A., Rodríguez-Pose, A. and Tomaney, J. (eds) *Handbook of Local and Regional Development*, 149–161. Routledge, London.

Bellini, N. and Hilpert, U. (eds) (2013). *Europe's Changing Geography: The Impact of Inter-Regional Networks*. Routledge, London.

Binz, C. and Truffer, B. (2017). Global innovation systems – A conceptual framework for innovation dynamics in transnational contexts. *Research Policy* 46:7, 1284–1298.

Blažek, J. and Kadlec, V. (2018). Knowledge bases, R&D structure and socio-economic and innovation performance of European regions. *Innovation: The European Journal of Social Science Research* 32:1, 26–47.

Bogliacino, F. and Pianta, M. (2010). Innovation and employment: A reinvestigation using revised Pavitt classes. *Research Policy* 39:6, 799–809.

Boschma, R. (2015). Towards an evolutionary perspective on regional resilience. *Regional Studies* 49, 733–751.

Braczyk, H., Cooke, P. and Heidenreich, M. (eds) (1998). *Regional Innovation Systems: The Role of Governances in a Globalized World*. Psychology Press, London.

Carayannis, E. and Campbell, D. (2009). 'Mode 3' and 'Quadruple Helix': Toward a 21st century fractal innovation ecosystem. *International Journal of Technology Management* 46:3–4, 201–234.

Carrincazeaux, C. and Coris, M. (2011). Proximity and innovation. In Cooke, P., Asheim, B., Boschma, R., Martin, R., Schwatz, D. and Tödtling, F. (eds) *Handbook of Regional Innovation and Growth*, 269–291. Edward Elgar, Cheltenham.

Chataway, J., Chux, D., Kanger, L., Ramirez, M., Schot, J. and Steinmueller, E. (2017). *Developing and Enacting Transformative Innovation Policy. A Comparative Study*. Transformative Innovation Policy Consortium, University of Sussex.

Chataway, J., Hanlin, R. and Kaplinsky, R. (2014). Inclusive innovation: An architecture for policy development. *Innovation and Development* 4:1, 33–54.

Chinsembu, K. (2015). Plants as antimalarial agents in Sub-Saharan Africa. *Acta Tropica* 152, 32–48.

Coenen, L., Asheim, B., Bugge, M. and Herstad, S. (2017). Advancing regional innovation systems: What does evolutionary economic geography bring to the policy table? *Environment and Planning C: Politics and Space* 35:4, 600–620.

Cohen, W. and Levinthal, D. (1990). Absorptive capacity: A new perspective on learning and innovation. *Administrative Science Quarterly* 35:1, 128–152.

Cooke, P. (1992). Regional innovation systems: Competitive regulation in the new Europe. *Geoforum* 23:3, 365–382.

Cozzens, S. and Kaplinsky, R. (2009). Innovation, poverty and inequality: Cause, coincidence, or co-evolution? In Lundvall, B., Joseph, K., Chaminade, C. and Vang, J. (eds) *Handbook of Innovation and Developing Countries: Building Domestic Capabilities in a Global Setting*, 57–82. Edward Elgar, Cheltenham.

Diercks, G., Larsen, H. and Steward, F. (2019). Transformative innovation policy: Addressing variety in an emerging policy paradigm. *Research Policy* 48:4, 880–894.

Doloreux, D. (2002). What we should know about regional systems of innovation. *Technology in Society* 24:3, 243–263.

Dutrénit, G. and Sutz, J. (eds) (2014). *National Innovation Systems, Social Inclusion and Development: The Latin American Experience*. Edward Elgar, Cheltenham.

Edquist, C. (2005). Systems of innovation: Perspectives and challenges. In Fagerberg, J., Mowery, D. and Nelson, R. (eds) *The Oxford Handbook of Innovation*, 181–208. Oxford University Press, Oxford.

Etzkowitz, H. and Dzisah, J. (2007). The triple helix of innovation: Towards a university-led development strategy for Africa. *ATDF Journal* 4:2, 3–10.

Etzkowitz, H. and Leydesdorff, L. (1995). The triple helix – university-industry-government relations: A laboratory for knowledge based economic development. *EASST Review* 14:1, 14–19.

Etzkowitz, H. and Leydesdorff, L. (2000). The dynamics of innovation: From national systems and "mode 2" to a triple helix of university – industry – government relations. *Research Policy* 29:2, 109–123.

Foray, D. (2014). *Smart Specialisation: Opportunities and Challenges for Regional Innovation Policy*. Routledge, London.

Freeman, C. (1987). *Technology Policy and Economic Performance: Lessons from Japan*. Frances Printer, London.

Frenken, K. (2017). A complexity-theoretic perspective on innovation policy. *Complexity, Innovation and Policy* 3:1, 35–47.

Garcia Haro, M., Martinez-Ruiz, M. and Martinez-Canas, R. (2014). Value co-creation process: Effects on the consumer and the company. *Export Journal of Marketing* 2, 68–81.

George, G., Howard-Grenville, J., Joshi, A. and Tihanyi, L. (2016). Understanding and tackling societal grand challenges through management research. *Academy of Management Journal* 59:6, 1880–1895.

George, G., McGahan, A. and Prabhu, J. (2012). Innovation for inclusive growth: Towards a theoretical framework and a research agenda. *Journal of Management Studies* 49:4, 661–683.

Gorjestani, N. (2004). Indigenous knowledge for development. In Twarog, S. and Kapoor, P. (eds) *Protecting and Promoting Traditional Knowledge: Systems, National Experiences and International Dimensions*, 265–272. United Nations, New York.

Hamel, K., Tong, B. and Hofer, M. (2019). Poverty in Africa is now falling – but not fast enough. *Brookings*, March 28.

Hartzenberg, T. (2011). Regional integration in Africa. WTO. *Economic Research and Statistics Division, Staff Working Paper ERSD* 2011–14.

Hautala, J. (2011). International academic knowledge creation and ba. A case study from Finland. *Knowledge Management Research & Practice* 9:1, 4–16.

Hautala, J. and Jauhiainen, J. (2014). Spatio-temporal aspects of knowledge creation. *Research Policy* 43, 655–668.

Hautala, J. and Jauhiainen, J. (2019). Creativity-related mobilities of peripherally located artists and scientists. *Geojournal* 84:2, 381–394.

Heeks, R., Foster, C. and Nugroho, Y. (2014). New models of inclusive innovation for development. *Innovation and Development* 4:2, 175–185.

Hooli, L. and Jauhiainen, J. (2017). Development aid 2.0 – towards innovation-centric development co-operation: The case of Finland in southern Africa. In Cunningham, P. and Cunningham, M. (eds) *IST-Africa 2017 Conference Proceedings*, 1–9. IIMC International Information Management Corporation, Windhoek, Namibia.

Hooli, L., Jauhiainen, J., Järvi, A., Nkonoki, E., Taajamaa, V. and Käyhkö, N. (2019). Contextualising innovation in Africa: Knowledge modes and actors in local innovation development. In *IST-Africa Week Conference (IST-Africa) Proceedings*, Nairobi, Kenya, 2019.

Ibert, O. (2007). Towards a geography of knowledge creation: The ambivalences between 'knowledge as an object' and 'knowing in practice'. *Regional Studies* 41:1, 103–114.

Isaksen, A. and Karlsen, J. (2010). Different modes of innovation and the challenge of connecting universities and industry: Case studies of two regional industries in Norway. *European Planning Studies* 18:12, 1993–2008.

Jauhiainen, J. and Hooli, L. (2017). Indigenous knowledge and developing countries' innovation systems. The case of Namibia. *International Journal of Innovation Studies* 1:1, 89–106.

Jauhiainen, J. and Suorsa, K. (2008). Triple helix in the periphery: The case of Multipolis in Northern Finland. *Cambridge Journal of Regions, Economy and Society* 1, 285–301.

Jensen, M., Johnson, B., Lorenz, E., Lundvall, B. and Lundvall, B. (2007). Forms of knowledge and modes of innovation. In Lundvall, B. (ed) *The Learning Economy and the Economics of Hope*, 155–182. Anthem Press, London.

Johnson, B. and Andersen, A. (2012). *Learning, Innovation and Inclusive Development*. Globelics-Global Network for Economics of Learning, Innovation, and Competence Building Systems. Department of Business and Management, Aalborg University, Aalborg.

Jurowetzki, R., Lema, R. and Lundvall, B. (2018). Combining innovation systems and global value chains for development: Towards a research agenda. *The European Journal of Development Research* 30:3, 364–388.

Kaplinsky, R. (2011). Schumacher meets Schumpeter: Appropriate technology below the radar. *Research Policy* 40:2, 193–203.

Knight Frank (2019). *The Wealth Report. The Global Perspective on Prime Property and Investment*. www.knightfrank.com/. Retrieved June 2019.

Kraemer-Mbula, E. and Wamae, W. (2010). Adapting the innovation systems framework to Sub-Saharan Africa. In Kraemer-Mbula, E. and Wamae, W. (eds) *Innovation and the Development Agenda*, 65–90. OECD, Paris.

Lee, N. (2016). Growth with inequality? The local consequences of innovation and creativity. In Sheamur, R., Carrincazeaux, C. and Doloreux, D. (eds) *Handbook on the Geographies of Innovation*, 419–431. Edward Elgar, Cheltenham.

Lundquist, K. and Trippl, M. (2013). Distance, proximity and types of cross-border innovation systems: A conceptual analysis. *Regional Studies* 47:3, 450–460.

Lundvall, B. (1992). *National Systems of Innovation: Towards a Theory of Innovation and Interactive Learning*. Pinter, London.

Lundvall, B. (2004). The economics of knowledge and learning. In Christensen, J. and Lundvall, B. (eds) *Product Innovation, Interactive Learning and Economic Performance*, 21–42. Emerald Group Publishing, Bingley.

Lundvall, B. (2016). Innovation as an interactive process: From user – producer interaction to the national systems of innovation. In Lundvall, B. (ed) *The Learning Economy and the Economics of Hope*, 61–84. Anthem Press, London.

Lundvall, B., Joseph, K., Chaminade, C. and Vang, J. (eds) (2009). *Handbook of Innovation and Developing Countries: Building Domestic Capabilities in a Global Setting*. Edward Elgar, Cheltenham.

Malecki, E. (2010). Global knowledge and creativity: New challenges for firms and regions. *Regional Studies* 44:8, 1033–1052.

Markusen, A. (2003). Fuzzy concepts, scanty evidence, policy distance: The case for rigour and policy relevance in critical regional studies. *Regional Studies* 37:6–7, 701–717.

McCann, P. and Ortega-Argilés, R. (2013). Modern regional innovation policy. *Cambridge Journal of Regions Economy and Society* 6:2, 187–216.

McCann, P. and Ortega-Argilés, R. (2015). Smart specialization, regional growth and applications to European Union cohesion policy. *Regional Studies* 49:8, 1291–1302.

Mêgnigbêto, E. (2013). Triple helix of university-industry-government relationships in West Africa. *Journal of Scientometric Research* 2:3, 214–222.

Moodysson, J. and Zukauskaite, E. (2014). Institutional conditions and *innovation* systems: On the impact of regional policy on firms in different sectors. *Regional Studies* 48:1, 127–138.

Moulaert, F. and Sekia, F. (2003). Territorial innovation models: A critical survey. *Regional Studies* 37:3, 289–302.

Muchie, M., Gammeltoft, P. and Lundvall, B. (eds) (2003). *Putting Africa First. The Making of African Innovation Systems.* Aalborg University Press, Aalborg.

Ndabeni, L., Rogerson, C. and Booyens, I. (2016). Innovation and local economic development policy in the global South: New South African perspectives. *Local Economy* 31:1–2, 299–311.

Nelson, R. (ed) (1993). *National Innovation Systems: A Comparative Analysis.* Oxford University Press, Oxford.

Nonaka, I. and Takeuchi, H. (1995). *The Knowledge-Creating Company: How Japanese Companies Create the Dynamics of Innovation.* Oxford University Press, Oxford.

Nonaka, I. and Toyama, R. (2015). The knowledge-creating theory revisited: Knowledge creation as a synthesizing process. In Edwards, J. (ed) *The Essentials of Knowledge Management*, 95–110. Springer, Berlin.

Oluwatobi, S., Efobi, U., Olurinola, I. and Alege, P. (2015). Innovation in Africa: Why institutions matter. *South African Journal of Economics* 83, 390–410.

Oyelaran-Oyeyinka, O. (2006). *Learning to Compete in African Industry: Institutions and Technology in Development.* Ashgate, Aldershot.

Oyelaran-Oyeyinka, O. (2014). The state and innovation policy in Africa. *African Journal of Science, Technology, Innovation and Development* 6:5, 481–496.

Pansera, M. and Martinez, F. (2017). Innovation for development and poverty reduction: An integrative literature review. *Journal of Management Development* 36:1, 2–13.

Parrilli, M. and Heras, H. (2016). STI and DUI innovation modes: Scientific-technological and context-specific nuances. *Research Policy* 45:4, 747–756.

Patra, S. and Muchie, M. (2018). Research and innovation in South African universities: From the triple helix's perspective. *Scientometrics* 116:1, 51–76.

Pavitt, K. (2002). Innovating routines in the business firm: What corporate tasks should they be accomplishing? *Industrial and Corporate Change* 11:1, 117–133.

Pavitt, K. (2005). Innovation processes. In Fagerberg, J., Mowery, D. and Nelson, R. (eds) *The Oxford Handbook of Innovation*, 86–114. Oxford University Press, New York.

Polanyi, M. (1966). *The Tacit Dimension.* University of Chicago, Chicago.

Ranga, M., Hoareau, C., Durazzi, N., Etzkowitz, H., Marcucci, P. and Usher, A. (2013). Study on university-business cooperation in the US. *LSE Research Online Documents On Economics* 55424.

Scerri, M. (2013). Modes of innovation and the prospects for economic integration in Africa. *Africa Insight* 43:3, 80–99.

Schillo, R. and Robinson, R. (2017). Inclusive innovation in developed countries: The who, what, why and how. *Technology Innovation Management Review* 7:7, 34–46.

Schot, J. and Steinmueller, W. (2018). Three frames for innovation policy: R&D, systems of innovation and transformative change. *Research Policy* 47:9, 1554–1567.

Schumpeter, J. (1934). *The Theory of Economic Development.* Harvard University Press, Cambridge, MA.

Shearmur, R., Carrincazeaux, C. and Doloreux, D. (eds) (2016). *Handbook on the Geographies of Innovation*. Edward Elgar, Cheltenham.

Soete, L. (2013). Is innovation always good? In Fagerberg, J., Martin, B. and Sloth Andersen, E. (eds) *Innovation Studies. Evolution and Future Challenges*, 134–146. Oxford University Press, Oxford.

Steffen, W., Richardson, K., Rockström, J., Cornell, S., Fetzer, I., Bennett, E. and Folke, C. (2015). Planetary boundaries: Guiding human development on a changing planet. *Science* 347:6223, 1259855.

Tödtling, F., Asheim, B. and Boschma, R. (2013). Knowledge sourcing, innovation and constructing advantage in regions of Europe. *European Urban and Regional Studies* 20:2, 161–169.

Tödtling, F. and Trippl, M. (2005). One size fits all? Towards a differentiated regional innovation policy approach. *Research Policy* 34:8, 1203–1219.

Trippl, M. (2010). Developing cross-border regional innovation systems: Key factors and challenges. *Tijdschrift voor Economische en Sociale Geografie* 101:2, 150–160.

United Nations (2015). *The 2030 Agenda for Sustainable Development*. United Nations, New York.

von Hippel, E. (1990). Task partitioning: An innovation process variable. *Research Policy* 19:5, 407–418.

von Hippel, E. (2015). Democratizing innovation: The evolving phenomenon of user innovation. *Journal für Betriebwirstschaft* 55:1, 63–78.

Watkins, A., Papaioannou, T., Mugwagwa, J. and Kale, D. (2015). National innovation systems and the intermediary role of industry associations in building institutional capacities for innovation in developing countries: A critical review of the literature. *Research Policy* 44:8, 1407–1418.

3

EVOLVEMENT OF DEVELOPMENT COOPERATION

Introduction

In this chapter we discuss development policy. In the contexts of developing countries, we combine the literature of development and innovation studies. The international development cooperation regime is undergoing rapid changes. The socio-economic development of many developing countries in the Global South and the increased critique of past approaches to development by the Global North has shifted the general focus of development cooperation from poverty reduction towards the "beyond aid agenda" with the objective of sustainable economic growth. Today, many aid-donating countries wish to benefit economically from their donated development aid by turning it into business-oriented programs and projects contributing to the economic growth of the aid-donating countries and their enterprises.

After the introduction, in Section 3.2, we start by defining the concept of development cooperation. Then, in Sections 3.3–3.8, we review the changes in development aid during the past decades. We pay particular attention to the most recent shifts in international development. We discuss innovations as a part of the recent development cooperation between the Global North and Global South. Although the history of development can be divided into regimes, as exemplified by in the Development Assistance Committee (OECD-DAC) countries and transnational organisations, the understanding of development has been far from unified among researchers, stakeholders and activists. For example, approaches relying on Marxist and liberal economist frameworks have triggered heated debates.

- Modernisation with rationality was a key goal of international development policies in the 20th century to achieve development with science, technology and democracy.
- In the early 21st century, the general focus of development cooperation has shifted towards global sustainable development goals and the beyond aid agenda, aiming to create reciprocal economic relations between the aid donor and the receiver.
- Broad participation of all related aid donating, receiving and cooperation stakeholders is needed in international development aid and policies.

Development cooperation and policy

The definition and the content of development have been in constant flux since its foundation due to the influence of global politics, economic policies and geopolitics (Hart, 2009). Development is a central but much-contested topic in many academic fields such as geography, politics, economics, education, psychology and development studies. Generally, development is considered social, economic and political progress, or a tool to accomplish such progress (Potter et al., 2018). Development is associated with efforts to improve the well-being of regions or countries, embracing people, structures and the environment within them, as well as tools, processes and strategies driving these changes (Willis & Kumar, 2009).

An inherent part of global development is the generic division of the world to the more developed Global North and the less-developed Global South (Potter et al., 2018). In general, the Global North includes Europe, parts of North America (i.e. the United States and Canada), parts of Asia (i.e. Israel, Japan, Singapore, South Korea and Taiwan [ROC]), and parts of Oceania (i.e. Australia and New Zealand). The Global South is made up of countries in Africa, Latin America and developing Asia, including the Middle East. Many countries, including the two most populous countries of China and India, are not part of either group.

This distinction of countries according to their development is more ideological than geographical. This dualistic North-South division has been heavily criticised as artificial and controversial. Nevertheless, it is still commonly used, even among more critical and progressive stakeholders who scrutinise hierarchies and exploitative processes of colonialism and post-colonialism that influence the interaction between the more developed northern and less-developed southern countries (Mawdsley, 2017). Although the concept of development is most often associated with the Global South, development occurs everywhere at various scales and diverse regions. For example, regions and people in the Global North face many similar challenges related to poverty, inequality, and global climate change. Hence, approaching development globally is essential.

As an active intervention, development means creating a better life for everyone by allowing most people to meet their basic needs such as food and water, good and healthy places to live, affordable services that fulfil the needs of people and the potentiality to develop oneself while being treated with respect and dignity (Peet & Hartwick, 2009). Development cooperation and development aid are often seen as synonyms, although cooperation is considered to be a more interactive and reciprocal process (Oden, 2010). Development aid is usually defined as one-way economic or material assistance to less-developed countries or regions. From this perspective, the recipients are not able to move from poverty without external assistance. Hence, development aid may also hinder empowerment and maintain dependencies. Such a development aid discourse was dominant until the end of the 20th century, when the cooperation discourse started to replace it. Afterwards, "development cooperation" replaced "aid" as the most common term. In development cooperation, recipients are regarded more as active stakeholders who, together with the donors, are involved in the planning of their development agenda. Development policy includes all incentives and strategic objectives guiding the development cooperation.

Traditionally, development actors have been divided into states, civil society and markets. However, new actors have emerged, and nowadays the actors are divided into five categories according to their financial aid and knowledge: members of OECD-DAC, non-DAC donors doing development cooperation, the private sector, global finance actors and NGOs (Banks & Hulme, 2014). For a long time, international development of the Global South was dominated by the OECD-DAC (see Box 3.1) together with transnational organisations, such as the UN, IMF and the World Bank, who were the most influential transnational aid institutions. As a particular form of international collaboration, the DAC distinguishes official development assistance (ODA) from other instruments of cooperation, such as military assistance and official credit export.

BOX 3.1   OFFICIAL DEVELOPMENT
ASSISTANCE (ODA)

OECD-DAC countries established the term official development assistance
(ODA) to describe loans, grants and other financial flows that meet the
requirements of the DAC and the OECD (Gore, 2013). Traditionally, ODA
has been almost synonymous with development cooperation, because the
DAC defines the terms for ODA and provides the largest share of development aid resources. Funded by the 30 official DAC countries, ODA
grew 30% between 2006 and 2016 before stagnating slightly in 2017
at $161 billion USD (OECD, 2018: 267). In 2017, Sub-Saharan Africa
remained the largest recipient region, receiving 22% of all bilateral ODA,
followed by South and Central Asia and the Middle East and North Africa,
both receiving 12% (Figure 3.1). In Africa, the largest per capita receivers
of ODA in 2016 were São Tomé and Príncipe ($242 USD per capita), Cabo
Verde ($215), Djibouti ($206), Liberia ($177) and South Sudan ($125),
while the average in Africa was $37 USD per capita (UNECA, 2018b: 85).

The share of the ODA relative to the DAC countries' gross national
incomes decreased slightly between 2015 and 2016. On the contrary,
the amount of humanitarian aid from the DAC countries doubled due to
investments related to global refugee management following the rapid
inflow of asylum seekers to Europe in 2015. The 2018 Development
Cooperation Report estimates that development assistance flows from
other donors continues to increase. For example, the South-South cooperation flows reached almost $8 billion USD in 2016. Yet ODA remains
the largest source of development aid internationally. This is despite the
changes and emerging actors in the new development landscape that
increasingly rupture this traditional structure. As Gore (2013) points out,
ODA as an external financing source has declined in the Global South,
whereas the significance of foreign direct investment (FDI) flows and
remittances has escalated. In Africa, the largest per capita inflows of FDI
in 2016 were Seychelles ($1,600 USD per capita), Angola ($556), Congo
($423), Gabon ($399), Mauritius ($274) and Cabo Verde ($227) when
the average in Africa was $48 USD per capita (UNECA, 2018b: 85).

The concept of "blended finance" responds to the complex development financing landscape. Established in the Addis Ababa Action Agenda
2015, the blended finance framework combines public finance with
private investment. The aim is to enhance private sector markets and

resources and to support the achievement of SDGs. These would be challenging to reach with the public financing from the ODA alone (United Nations, 2018: 102–103). The Financing for Development report by the UN in 2018 argues that blended financing improves the effectiveness of assistance, even though it calls for other actions as well. According to the report, the donor countries should aim to provide untied development aid (i.e. aid that has no conditions on the recipient procuring goods and services from the donor countries). A fifth (19%) of the ODA was tied in 2016 (United Nations, 2018: 106). However, as Quadir (2013) notes, the increase in the effectiveness of assistance also requires focussing on what constructive needs and social change dynamics of the recipient countries bring to changes in the Global South.

Net ODA by regions, 2016 (million USD)

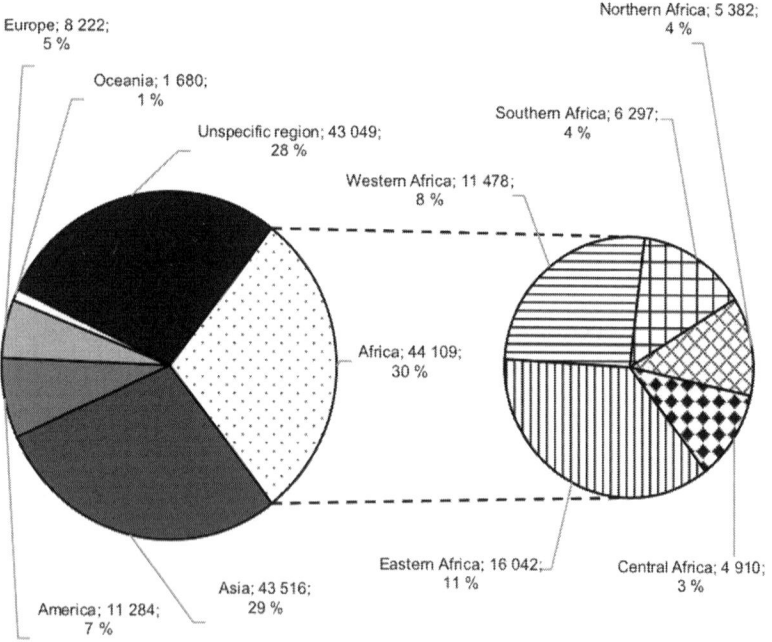

FIGURE 3.1 Development aid: Financial flows of OECD-ODA

Source: Modified from OECD (2019)

The definition and the public image of development cooperation are often perceived only from altruistic perspectives. This means that development cooperation aims to contribute to local communities' livelihoods and well-being by positively influencing larger development dynamics such as poverty and inequality (Headey, 2008). However, the seemingly altruistic motivations behind development cooperation are always intertwined with other motivations and unequal power relations (Chambers, 2008). The self-interest of donors and expectations for reciprocity have always been part of international development (Alesina & Dollar, 2000; Temple, 2010). For example, during the 1980s, the United States also granted loans to Global South countries to strengthen the value of the USD. In addition, several donor countries have supported specific industrial sectors to create new markets for their products, obtain inexpensive labour for manufacturing or have access to raw materials. More recently, the emergence of innovations in development cooperation has also increased and enhanced the absorptive capacity of the Global South countries and the economic growth of aid-receiving countries, and by achieving this it creates new markets for the donor countries' technology companies (Ylönen, 2012).

Several researchers have criticised that the aforementioned self-interest behind the current development cooperation widens inequalities among the populations of the aid-receiving countries (Power, 2003; Mawdsley, 2015). As part of the assistance regime, the donors especially reward the recipients' actions in selected fields (Hoeffler & Outram, 2011); for example, to support the donor countries in their transnational organisations, accept more refugees in the recipient countries or make positive environmental actions there. Moreover, development cooperation is driven by visions of international politics and contains moral presumptions. For example, during the Cold War period, development aid was explicitly connected to geopolitics with the aim of preventing the domino effect across the Third World, whether toward alliances with the Soviet Union or the United States. Geopolitical interests also continue in the current development cooperation even though many former political blocs have disappeared.

Modernisation

Contemporary development approaches are rooted in the 18th century, emphasising transformation driven by science, technology and culture in Europe (Power, 2003). With rational interventions supported by science and technology, the world could be changed for the better. Its roots are embedded in the history of the Enlightenment era and colonialism.

In practice, development was not then considered from international over-arching perspectives. Instead, states focussed on fostering their own national development. Colonialism prevailed in the relations between the more developed mother countries and the less-developed areas, of which some were direct colonies of the mother countries. In fact, since the 15th century, development in Europe and the United States was promoted by their colonial abuse of the material and social resources of the less-developed areas, including enslavement of their inhabitants, especially in Africa.

The early international development regime from post-World War II until the 1980s was inspired by the modernisation theory (Rist, 2007). Development aid was born after the Allied forces won World War II and the President of the United States, Harry S. Truman (1884–1972), offered help to less fortunate nations (Moyo, 2009). He launched the four-point programme in which modernist development was regarded as structural transformation from traditional agricultural societies towards Western lifestyles, industrialism, urbanisation and economic policies to accomplish growth. The development projects included financial and material investments and spreading knowhow, mainly to growing urban centres (Rocha, 2013). The assumption was that from these urban centres, the benefits of economic growth would further advance the living standards and advantages of the entire population, as the growth was expected to trickle down towards more peripheral regions.

One of the key actors behind the modernisation theory was economic historian Walt W. Rostow (1916–2003). He presented a theory of universal development stages which all societies go through. According to this theory, global consumption will constantly increase. Eventually, the whole world becomes integrated in the capitalist system and ideology. During their development, all societies go through the same five basic stages, although at varying lengths. The stages are as follows: traditional society, preconditions for developmental take-off, take-off with technology and investments, drive towards mature development, and high industrial mass consumption (Rostow, 1960; Peet & Hartwick, 2009: 127–129). Rostow believed that the countries in the Global South could only achieve development via external assistance and technological development.

Another major figure in modernisation theory, whose ideas were utilised in development theories, was the British economist John Maynard Keynes (1883–1946). He is associated with the rise of modern liberalism. Distinctively from Rostow's model, Keynes argued that economic progress requires a strong national state with strong fiscal and monetary policies. A state can support development by lowering interest rates through its central bank, thus fostering investment and economic turnover. The growing governmental spending on both fixed (machines, factories and technology) and human capital (education and skills of labour force) would boost the economy when needed (Peet & Hartwick, 2009: 56–59).

The modernisation theory later became a subject of criticism and was mostly rejected as the main development approach. One of the most criticised aspects was the idea that technology, industrial development or economic growth can be transferred from one geographical location to another, regardless of the differences in their local socio-economic contexts. According to the critique, development in the Global South cannot follow the same path as the earlier industrialised countries in Europe. This is an aspect that is again topical, as current development cooperation that focuses on innovations relies on some basic ideas of modernisation (see Chapter 4).

Modernisation has also been criticised due to its Eurocentric approach and promotion of homogenous sociocultural values. From this perspective, the Global South is conceived to be backward and the indigenous habits and traditions there hinder development. This has been used as an excuse to wipe away indigenous cultures during modernisation processes. Modernisation was an especially contradictory approach in Africa, where a long colonial history had forcefully legitimised the supremacy of the Western civilisation. After the wave of independence in the early 1960s, many African countries struggled with their attempts to balance modernisation of their economies with respecting their indigenous cultures. Furthermore, the idea of economic trickle-down was criticised as urban-rural development bias and inequalities continued to grow (Langan, 2011). Despite this criticism, modernisation never disappeared from the development agenda, but transformed into other approaches, as later explained.

Despite the hegemony of the economic growth approach, by the 1970s, the understanding of development evolved to become more multidimensional. Objectives such as poverty reduction focussing on enhancing education, providing better social services and improving nutrition emerged in the donors' agenda (Moyo, 2009). In development studies, the structuralist and Marxist explanations became increasingly common (Willis & Kumar, 2009). According to the structuralist approach, economic, social and political structures prevent countries from achieving development. On the contrary, the Marxist approach conceived that capitalism combined with colonialism is the main driver of inequality. It argued that the Global North achieves its progress by exploiting the countries of the Global South and their resources. Generally, this is called the dependency theory.

Washington Consensus and the structural adjustments

In the 1980s, development aid and, more broadly, the notion of development were still mainly determined by the focus on economic growth as measured by the growth of the aid-receiving country's GDP (Rocha, 2013). However, the difference from the previous development policy was the changing role of the state. The Keynesian state-supported economic growth model was regarded as offering too few opportunities for market-driven development (Potter et al., 2018). This was evidenced by the economic recession (1979–1983) caused by the rise of energy prices (the so-called second oil crisis) that led the countries in the Global South into a debt crisis, meaning that several countries could no longer cope with their debts (Soederberg, 2004). During the debt crises, countries spent a large part of their GDP on paying interest and debts instead of investing in necessary productive infrastructure, education and public services.

This epoch of development was commonly called the Washington Consensus or the neoliberalist consensus (Overton & Murray, 2016). The focus was on creating macroeconomic stable growth with the opening of the economy and the expansion of market forces. Market-led policies and privatisation minimised the role of the state and opened the economy for trade and investments. International financial institutions, in particular the World Bank and the IMF, encouraged less-developed countries to follow the so-called "laissez-faire" policies and seek solutions from the market economy (Gore, 2000). The idea was that free market development supported by FDI would create economic growth and prosperity. The World Bank and the IMF presented structural adjustment programs as a solution to the debt crisis and to stabilise and adjust the countries' economies. This resulted in the privatisation wave of several public services in the Global South (Gibbon & Schulpen, 2002).

During this regime, multinational companies began to relocate their labour-intensive industries to developing countries. The motivation was to lower production costs and gain higher economic returns. The development aid shifted from domestic development to broader international development assistance and cooperation led by international financial institutions (Gore, 2000). Private companies were recognised as key development actors for the first time. This caused the decline of multilateral and bilateral development assistance programmes.

The short-term objectives and international financial institutions pressured development projects for quick financial results. Unlike in the Keynesian model, neoliberalism emphasised individuals' ability to cope on their own. According to critics, the neoliberal ideology increased inequality and concentrated power in the hands of a few (Harvey, 2007). Due to structural reforms, many people dependent on employment in public services were impoverished and income inequalities increased. In addition, multinational companies started to invest in the Global South to benefit from inexpensive unorganised labour and weak control of working and environmental conditions. Raw materials from the Global South were exported to the Global North where a larger value was added to them. Thus, the 1980s is often called "the lost decade of development" (Potter et al., 2018: 23). The traces of neoliberalism are still visible, and many development projects still seek short-term development gains. In fact, neoliberal development projects were often led by foreign experts without proper understanding or interest in the local socio-economic context.

Post-Washington Consensus

Although neoliberal ideology was still strong, in the 1990s the development paradigm was transformed again. Arguments increased that the top-down policy interventions had failed in their envisioned economic development delivery in resource-limited settings in the Global South (George et al., 2012). This time, there was a shift from mere economic growth orientation toward more diverse development approaches. In the late 1990s, the economies of the East Asian countries fell into crisis. This resulted in the revision of the development standards of the World Bank and the IMF. The simple structural adjustments turned out to be unsuccessful as countries from the Global South were unable to pay back their debts. As a result, states and international institutions once again recognised the need to intervene in the global economy.

In the 1990s, empowerment from the post-Washington Consensus and the eradication of poverty became the main goals of international development (Soederberg, 2004). The less-developed countries should help themselves by conducting reforms correcting market limitations, enhancing national fiscal and financial institutions and promoting good governance. Development required a balanced role between the state and the market, and countries should implement contextualised solutions to achieve this. The development cooperation itself remained a top-down exercise but the perspectives of development strategies better acknowledged the contextual differences with the long-term goals. Economic development was seen as a qualitative change that also contributed to job creation (Rocha, 2013). Economic development was still seen as the best way to raise equality among the inhabitants of the Global South (Schulpen & Gibbon, 2002; Langan, 2011).

Although the idea of development shifted from admiring and following the Western countries to more plural approaches, people's living conditions did not significantly improve, the state of the environment continued to deteriorate, and inequality among populations in the Global South increased (Rocha, 2013). One reason for the growing income disparities was the inequality of world trade. The countries of the Global South had to spend more money due to trade tariffs and restrictions than they received in the form of international development aid from the Global North. The development actors became aware that in addition to economic resources, other measures to overcome poverty and inequality are also needed (Willis, 2011).

Human and sustainable development

While the Washington Consensus was a globally hegemonic international development ideology in the 1970s and 1980s, a new development ideology emphasising human development started to emerge in the 1980s (Gore, 2000). Various global campaigns, such as Live Aid in 1985, gained popularity, raised awareness of development challenges and the need of humanitarian concern in foreign policy, and shaped public opinion about neoliberalism.

Initially, the post-development approach was born as a grassroots movement in the 1960s. It opposed the Washington Consensus and, more broadly, also capitalism. These opponents saw colonialism, structural adjustments and unfair trade rules set by the Global North as the causes of poverty. In particular, the dependency theories from Latin America criticised the unequal power relations between the more developed capitalist states in the Global North and the less-developed countries elsewhere (Escobar, 1995; Silvey, 2010). The post-development approach became attached to post-colonial theory, which aimed to eradicate post-colonial dependencies that aggravated the cycle of poverty by strengthening the role of the state as an engine of development (Willis & Kumar, 2009).

The post-development school of thought criticised the Western-dominated, top-down development programmes and the globalist development hegemony with its narrow universal economy-oriented definition and development projects (e.g. Escobar, 1995; Rahnema, 1997). The post-development approach sought to improve the living conditions of the poor and appreciate local communities and cultures. It acknowledged the importance of local people's involvement and participation as active actors in their own development, rather than trusting foreign experts with their top-down policies and implementation of development projects (Rocha, 2013). However, the post-development theory and other grassroots movements have also been criticised for their remarkably radical arguments without appropriate solutions or practicality.

Later, the dependence theories and other grassroots movements were strongly influenced by the writings of Amartya Sen (b. 1933), the winner of the Nobel Memorial Prize in Economic Sciences in 1998, about development as freedom. He considered development as an individual human freedom to choose the lifestyle that appeals to oneself (Sen, 1997b). According to Sen, development is based on human capital and human capability. Therefore, poverty is a lack of freedoms and capabilities. It is not so much what people own, but what they are able to do or experience. As active stakeholders, individuals can influence development through their abilities, knowledge and efforts. Only by removing constraints on individuals' abilities and capacities is it possible to increase human well-being. In particular, Sen (1997a) emphasised that education, training and learning, and health contribute positively to economic prosperity

Development based on grassroots-level empowerment was also called "community development", in which community members work together to solve common development problems (McEwan et al., 2017). However, community development is criticised for its vague definition of community. In addition, the extended focus on community may suppress individuals' rights. Participation only extends the power of the Global North by not giving proper agency to local people (Langan, 2011: 89).

These novel ideas of participation also entered into development cooperation. The main objective was to involve communities in the design and implementation of development projects so that they would better fit into local contexts (Gore, 2000). At the same time, development began to be understood more multi-dimensionally by considering not only human well-being but also environmental issues. Environmental issues later became among of the most important objectives of international development (Potter et al., 2018: 23). The importance of multinational institutions in development cooperation was also emphasised by grassroots organisations (Gore, 2000: 791–793).

Development also needs to be measured to know the impact of interventions. According to Sen (1997b), human development can be measured either directly as an impact on people's daily lives or indirectly with broader benefits for society (e.g. through economic growth). Often development was measured indirectly, but sometimes both indirect and direct measurements were used simultaneously. The measurement of development must recognise personal experiences, which supports the integration of qualitative approaches in assessing development.

In the 1990s, the UN launched the Human Development Index (HDI) to measure development. Education and health were taken into account in addition to GDP (in the African context, see Figure 3.2). Since then, the HDI has been widely adopted and the UN Development Program (UNDP)

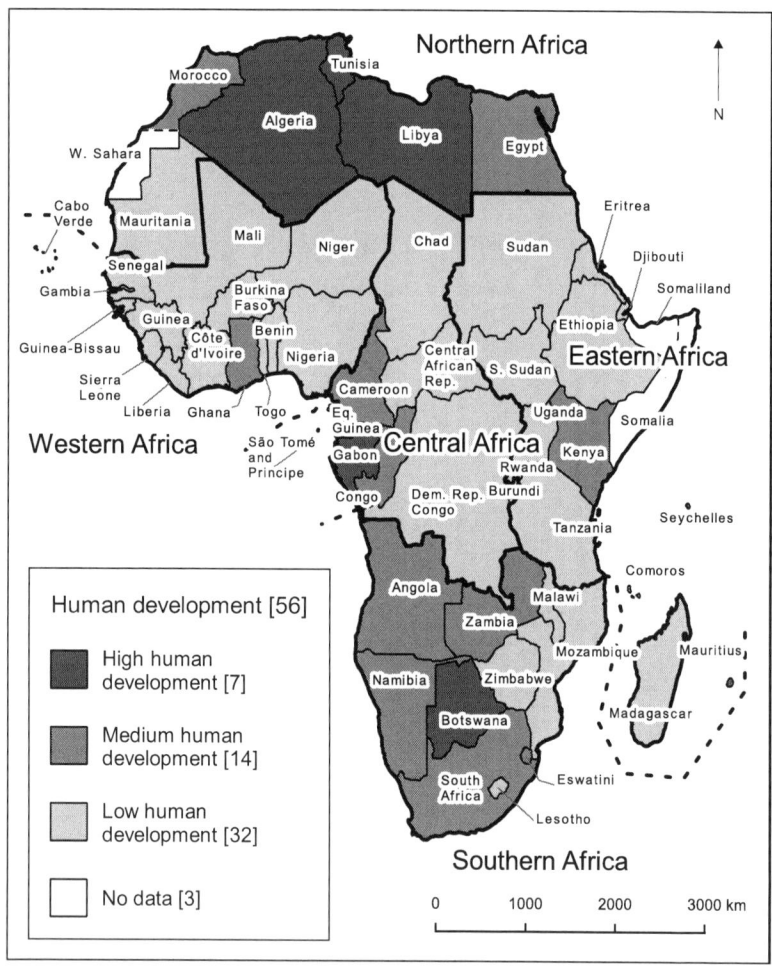

FIGURE 3.2 Human development in Africa based on HDI

Source: Modified from UNDP (2018)

annually publishes its Human Development Report, which divides countries into different categories according to their HDIs (UNDP, 2018). The writings of Sen (1997a, 1997b) and scholars resisting broader neo-liberalism had a major impact on development research. However, a more extended institutional change towards human development only occurred in the mid-1990s. Then human rights became the focus of development approaches (Rocha, 2013: 17). Poverty reduction became the main objective of international development cooperation.

The development as freedom approach has also been criticised. In particular, Corbridge (2002) criticises the inappropriate acknowledgement of possible conflicts between the freedoms of individuals and freedoms of communities or groups. In addition, according to the critique, Sen's writings did not sufficiently regard the connection between political regimes and economic development. Economic development does not automatically increase individual freedoms. In addition, power relations must always be considered.

United Nations development goals

In 2000, the UN launched its Millennium Development Goals (MDGs). MDGs set eight goals as the main objectives of global development to be reached by 2015. Each goal had a special target and temporal deadline for achieving the target. The MDGs were as follows (United Nations, 2000):

1   Eradicate extreme poverty and hunger;
2   Achieve universal primary education;
3   Promote gender equality and empower women;
4   Reduce child mortality;
5   Improve maternal health;
6   Combat HIV/AIDS, malaria and other diseases;
7   Ensure environmental sustainability;
8   Develop a global partnership for development.

These MDGs were the first time when the reduction of poverty and inequality were formed as the central principles of international development aid (Murray & Overton, 2016). They were supported with other extended related objectives such as good governance, economic growth, sustainable development, gender empowerment, education, health, inclusive finance and social welfare (Roy, 2010). The MDGs were an important milestone, as they were the first truly global development goals. They increased the role of international institutions through transnational cooperation (Fukuda-Parr et al., 2013). The MDGs also delegated more responsibility to the aid-receiving countries of their own development instead of top-down agendas that had been designed elsewhere and had become hard to adopt by the recipient countries in different socio-economic contexts (Mawdsley et al., 2014). Furthermore, in 2005, the principles of implementation and monitoring of development assistance were agreed upon with the Paris declaration that provided more tangible rules and goals for international development assistance. Later, this was strengthened and its implementation deepened with the Accra Action Agenda in 2008.

Despite many improvements that the MDGs brought to development assistance, they were criticised due to the lack of justification and analyses behind the chosen targets, their modesty and their focus on only the countries of the Global South (Saith, 2006). Furthermore, according to the criticism, the focus of the international declarations remained more on the agenda than on the implementation. Eventually, the achievement of goals progressed very unevenly and many large goals were not accomplished at all (Deacon, 2016).

A major change in the global recognition of development issues took place in 2015. Then the UN launched new Sustainable Development Goals (SDGs) that set the guidelines for international development cooperation until 2030. These were synthesised with a chart that explained the goals with a short text and symbols (see Figure 3.3). The 17 SDGs were as follows (United Nations, 2015):

1   No poverty;
2   Zero hunger;
3   Good health and well-being;
4   Quality education;
5   Gender equality;
6   Clean water and sanitation;
7   Affordable and clean energy;
8   Decent work and economic growth;
9   Industry, innovation and infrastructure;
10 Reduced inequalities;
11 Sustainable cities and communities;
12 Responsible consumption and production;
13 Climate action;
14 Life below water;
15 Life on land;
16 Peace, justice and strong institutions;
17 Partnerships for the goals.

FIGURE 3.3   The United Nations' Sustainable Development Goals

Source: Modified from United Nations (2015)

The aims of the SDGs differed from the MDGs by definition, scale, actors and focus. The aims of the SDGs are to better acknowledge various dimensions of sustainable development, such as ecological, social and economic sustainability (Rocha, 2013). Special emphasis was placed on ecological sustainability that was not efficiently addressed in the MDGs. New goals also provided greater visibility for the elimination of gender, ethnicity, age and disability discrimination (Deacon, 2016). The private sector had an important role in creating SDGs, and multinationals were especially able to attend and speak at the UN meetings (Kindornay & Reilly-King, 2013).

Beyond aid regime

The turbulence around development cooperation has continued during the 2010s. The development regime is turning away from poverty reduction towards the so-called "beyond aid agenda" (Janus et al., 2015; Mawdsley, 2017). Several researchers argue that the latest turning point in international development aid occurred after the 2011 Busan High Level Forum on Aid Effectiveness. Later, the UN SDGs turned the focus of international development towards development effectiveness (Gore, 2013; Quadir, 2013; de Renzio & Seifert, 2014; Murray & Overton, 2016).

The beyond aid agenda includes new, more conservative approaches that revive some principles from the previous development regimes of neoliberalism and modernisation (Sheppard & Leitner, 2010). The central objective of international development assistance is economic development, again replacing the direct target of poverty alleviation. Growth should be led by the private sector, which is regarded as the main engine by which to implement neoliberal policies (Mawdsley, 2017). The beyond aid agenda has brought new actors, finance, regulations and knowledge to international development (Janus et al., 2015). However, perhaps the most significant shift has been the reorientation of the traditional connection between the Global North and the Global South axes and the increase of South-South development cooperation.

The beyond aid agenda considers development through a relationship where inclusive and result-oriented partnership is reinforced to support local leadership, responsibility and capability (Michel, 2016: 20). Development assistance cannot function in isolation. It needs to be decoupled from trade, science, technology and innovation, immigration and environment policies (Mawdsley et al., 2018). The new development agenda integrates all kinds of international cooperation that may affect development including, for example, private sector development, smart aid, financialisation, technology development, innovations, competitive bidding for aid, growth-oriented entrepreneurship and mutual business interests between the donor and the receiver (Villanger, 2016). Furthermore, in addition to monetary support, expertise, experience and knowledge are also becoming increasingly important aspects in development policy.

The previous research discovered that although the beyond aid agenda is not based on a single ideology, the trajectory of international development assistance has followed the same path in various DAC countries, like the United Kingdom, the Netherlands, Canada, Australia and New Zealand (Murray & Overton, 2016). Many countries have restructured their public sector international development units and decoupled their ministries for development from their ministries of foreign trade.

This development aid regime shift results from the interplay between various events, actors and trends. First, the economic progress of several developing countries in the Global South has been rapid, and therefore many of them are becoming less dependent on donors' aid. The increased difference between countries and regions has made the uniform concept of the Global South gradually redundant. Many previous developing countries have already reached the status of middle-income countries. The rise of the Global South has increased the will of many previously developing countries to participate in development, and South-South cooperation is becoming increasingly popular. New development partners such as China and India are at the forefront, but countries like Turkey, Brazil, South Africa and Saudi Arabia have also brought new norms, ideas and practices to international development cooperation. This had ended the full dominance of the OECD-DAC to define what development is and what instruments should be used to achieve it (Hansen, 2018). The non-DAC donor countries' definition of development assistance is not universal. Development assistance is decoupled from other investments and bilateral deals. This makes it difficult to measure the amount invested in development.

The donors from the Global South, many of whom are previous ODA receivers and have followed different evolutionary socio-economic paths, are positioned differently in the global power structures, and have different understandings of how development should be arranged compared with the OECD-DAC (Mawdsley, 2017). That is why the development agenda of the non-DAC actors is formed by different domestic capacities and agendas. The development assistance by the non-DAC actors is a very diverse geo-cultural dynamic of giving and receiving (Bayly, 2009). Although the general development assistance from the Global South donors is connected to humanitarian and technical cooperation in health, education and welfare, many non-DAC donors also emphasise larger modernisation projects of transport and energy infrastructure, agriculture and manufacturing (de Renzio & Seifert, 2014). Opposed to the customary vertical North-South relations, South-South cooperation is more horizontal, targets reciprocal benefits and is unbiased towards domestic affairs.

Second, the post-global financial crisis (the economic downturn that started in 2007/2008) era and the populistic political movements in the Global North have raised critical tones towards the neostructuralist development assistance focussing mostly on poverty reduction (Banks & Hulme, 2014). Political and public pressure has led to the redefinition and reduction of donor countries' funding for less-developed countries. The reciprocity of development assistance is increasingly required. The private sector has again become the major vector and partner of development cooperation (Tomlinson, 2012; Di Bella et al., 2013; Blowfield & Dolan, 2014). Private enterprises, investments and trade are gaining a more substantial role in newly legitimised modalities of development financing (Severino & Ray, 2009; Griffiths, 2013). It is expected that the private sector will nurture investments, contribute to market efficiencies and self-regulating markets, improve income levels by generating new and better-quality jobs, and enhance tax revenues (Jeppesen, 2005; Di Bella et al., 2013). Furthermore, new inclusive innovations are expected to solve several poverty-related challenges. The role of the private sector in development assistance is discussed in more detail in Chapter 5.

Even though development aid has always benefitted the donor as well as the recipient, the beyond aid regime explicitly focuses on the interests of the donors and the private sector. Such an approach has similarities with the modernisation era of the 1950s–1960s, when donors invested substantially in infrastructure development (Murray & Overton, 2016; Mawdsley, 2017). China has vigorously funded infrastructure and energy projects to advance the means for the Chinese private sector and state-led enterprises to thrive in the Global South, including in Africa (see Chapter 8). In addition, other donor countries utilise their native manufacturers and suppliers. They thus direct a portion of the aid flows back to the donor country. This consequently reinforces the tied aid agreements, which the OECD encourages them to abandon. Development aid is seen also as a potential for increasing trade from donor countries to Africa. The DFID (2017), the United Kingdom's development agency, emphasised in 2017 the need to revitalise the trade deals with the less-developed countries following Brexit, the United Kingdom's (potential) departure from the EU.

Third, although there have been concerns about the widening digital divide between the developed and developing countries, the access to and

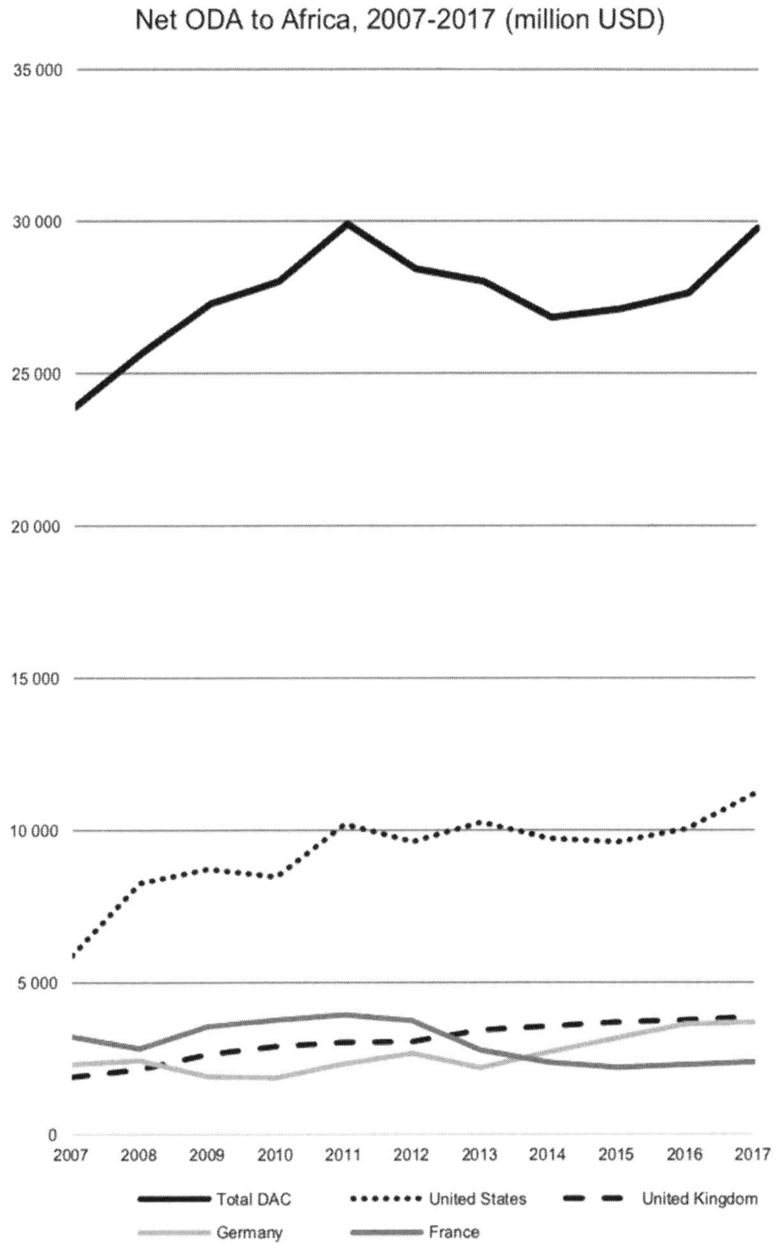

FIGURE 3.4 Official development aid to Africa. Global and country specific receivers of gross bilateral ODA from Africa, 2007–2017

Source: Modified from OECD (2019)

availability of mobile technologies have enhanced local populations' access to services, markets and knowledge in the Global South (Taylor, 2016). Developing countries are no longer solely end users of new technologies, but new innovations are also developed there (see Chapter 5). The growing young middle-class population and so-called pro-poor technologies result in several actors from the private sector searching for technological opportunities to refine their existing innovations and identify new needs for innovation in the Global South. The development of a knowledge society, digitalisation, opportunities for technology to solve development issues and positive examples of innovation systems as the engine of socio-economic development have increased the move toward innovation-focussed development cooperation. Chapter 4 discusses the role of innovation in development cooperation in more detail.

Conclusions

During the past 50 years, development has been understood from various perspectives and many different frameworks have been used to intervene in development. The positions of the state, the market and the people whom development is meant to change have varied substantially. International development assistance has oscillated between state-led and market-led approaches. Sometimes the people in the Global South have had something to say on the development initiatives, but other times development strategies, plans and projects designed in the Global North have just landed over them.

Mawdsley (2017) contemplates whether the current beyond aid era in international development assistance merely provides another narrative of "creative destruction" to the development paradigm, or whether this new development approach could produce long-lasting results with positive outcomes on people's livelihoods and well-being in the Global South. In addition, certain countries foster their geopolitical interests through development cooperation.

Does the economy or the people come first in development assistance programs, and which people are to benefit from such assistance? From the humanitarian perspective, the focus of development should remain on marginalised groups and inclusive social institutions to achieve a positive change rather than favouring markets over people. As Banks and Hulme (2014: 189) argue, to generate prosperous developmental outcomes, one should move towards an "economic and social process that also facilitates redistributions of power, representation and accountability, and more inclusive social, economic and political institutions".

Despite the growing importance of the private sector and the global market, the role of the state cannot be overlooked. In many cases, governments are involved in political, social and economic institutions and allocate there the resources of the state. This premise also supports the foundation of the transformative development cooperation framework (see Chapters 2 and 5). However, it remains crucial to comprehend development thoroughly (i.e. over history, objectives, actors and institutions) to contemplate new approaches and solutions to the development problems occurring today.

• Globally practiced sustainable development goals and the beyond aid agenda underline the importance of innovations for development.
• There are similarities between current development aid cooperation and the past modernisation agenda of development.
• States play still an important role in development between the Global North and the Global South, including Africa.

Discussion questions

• What is the beyond aid development cooperation?
• What similarities and differences exist between the modernisation and the beyond aid approaches in development assistance in Africa?
• Discuss how beyond aid development cooperation could take into account the interests of the different segments of population in Africa, as well as the interests of the Global North donor countries.

References

Alesina, A. and Dollar, D. (2000). Who gives foreign aid to whom and why? *Journal of Economic Growth* 5:1, 33–63.

Banks, N. and Hulme, D. (2014). New development alternatives or business as usual with a new face? The transformative potential of new actors and alliances in development. *Third World Quarterly* 35:1, 181–195.

Bayly, S. (2009). Vietnamese narratives of tradition, exchange and friendship in the worlds of the global socialist ecumene. In West, H. and Raman, P. (eds) *Enduring Socialism. Explorations of Revolution and Transformation, Restoration & Continuation,* 125–147. Berghahn Books, Oxford.

Blowfield, M. and Dolan, C. (2014). Business as a development agent: Evidence of possibility and improbability. *Third World Quarterly* 35:1, 22–42.

Chambers, R. (2008). *Revolutions in Development Inquiry. Institute of Development Studies.* Earthscan, London.

Corbridge, S. (2002). Development as freedom: The spaces of Amartya Sen. *Progress in Development Studies* 2:3, 183–217.

De Renzio, P. and Seifert, J. (2014). South – South cooperation and the future of development assistance: Mapping actors and options. *Third World Quarterly* 35:10, 1860–1875.

Deacon, B. (2016). Assessing the SDGs from the point of view of global social governance. *Journal of International and Comparative Social Policy* 32:2, 116–130.

DFID (Department for International Development) (2017). *Economic Development Strategy: Prosperity, Poverty and Meeting Global Challenges*. DFID, London.

Di Bella, J., Grant, A., Kindornay, S. and Tissot, S. (2013). *The Private Sector and Development: Key Concepts*. North-South Institute, Ottawa.

Escobar, A. (1995). Conclusion: Imagining a post-development era. In Crush, J. (ed) *Power of Development*, 211–227. Routledge, London.

Fukuda-Parr, S., Greenstein, J. and Stewart, D. (2013). How should MDG success and failure be judged: Faster progress or achieving the targets? *World Development* 41, 19–30.

George, G., McGahan, A. and Prabhu, J. (2012). Innovation for inclusive growth: Towards a theoretical framework and a research agenda. *Journal of Management Studies* 49:4, 661–683.

Gibbon, P. and Schulpen, L. (2002). Comparative appraisal of multilateral and bilateral approaches to financing private sector development in developing countries. *WIDER Discussion Paper* 112. Helsinki.

Gore, C. (2000). The rise and fall of the Washington consensus as a paradigm for developing countries. *World Development* 28:5, 789–804.

Gore, C. (2013). *Regions in Question: Space, Development Theory and Regional Policy*. Routledge, London.

Griffiths, M. (2013). *Realism, Idealism and International Politics: A Reinterpretation*. Routledge, London.

Hansen, T. (2018). *Wages of Violence: Naming and Identity in Postcolonial Bombay*. Princeton University Press, Princeton.

Hart, G. (2010). D/developments after the Meltdown. *Antipode* 41, 117–141.

Harvey, D. (2007). Neoliberalism as creative destruction. *The Annals of the American Academy of Political and Social Science* 610, 22–44.

Headey, D. (2008). Geopolitics and the effect of foreign aid on economic growth: 1970–2001. *Journal of International Development* 20:2, 161–180.

Hoeffler, A. and Outram, V. (2011). Need, merit, or self-interest – what determines the allocation of aid? *Review of Development Economics* 15:2, 237–250.

Janus, H., Klingebiel, S. and Paulo, S. (2015). Beyond aid: A conceptual perspective on the transformation of development cooperation. *Journal of International Development* 27:2, 155–169.

Jeppesen, S. (2005). Enhancing competitiveness and securing equitable development: Can small, micro, and medium-sized enterprises (SMEs) do the trick? *Development in Practice* 15:3–4, 463–474.

Kindornay, S. and Reilly-King, F. (2013). *Investing in the Business of Development*. www.nsi-ins.ca/wp-content/uploads/2013/01/2012-The-Business-of-Development.pdf/. Retrieved June 2019.

Langan, M. (2011). Private sector development as poverty and strategic discourse: PSD in the political economy of EU-Africa trade relation. *Journal of Modern African Studies* 49:1, 83–113.

Mawdsley, E. (2015). DFID, the private sector, and the recentring of an economic growth agenda in international development. *Global Society* 29:3, 339–358.

Mawdsley, E. (2017). Development geography 1: Cooperation, competition and convergence between 'North' and 'South'. *Progress in Human Geography* 41:1, 108–117.

Mawdsley, E., Murray, W., Overton, J., Scheyvens, R. and Banks, G. (2018). Exporting stimulus and "shared prosperity": Reinventing foreign aid for a retroliberal era. *Development Policy Review* 36, O25–O43.

Mawdsley, E., Savage, L. and Kim, S. (2014). A 'post-aid world'? Paradigm shift in foreign aid and development cooperation at the 2011 Busan High Level Forum. *The Geographical Journal* 180:1, 27–38.

McEwan, C., Mawdsley, E., Banks, G. and Scheyvens, R. (2017). Enrolling the private sector in community development: Magic bullet or sleight of hand? *Development and Change* 48:1, 28–53.

Michel, J. (2016). *Beyond Aid: The Integration of Sustainable Development in a Coherent International Agenda*. Rowman & Littlefield, London.

Moyo, D. (2009). *Dead Aid: Why Aid Is Not Working and How There Is a Better Way for Africa*. Macmillan, London.

Murray, W. and Overton, J. (2016). Retroliberalism and the new aid regime of the 2010s. *Progress in Development Studies* 16:3, 244–260.

Oden, B. (2010). The UN and development: From aid to cooperation. *Forum for Development Studies* 37:2, 269–279.

OECD (Organization for Economic Co-operation and Development) (2018). *Development Co-operation Report 2018: Joining Forces to Leave No One Behind*. 471 pp. OECD, Paris.

OECD (Organization for Economic Co-operation and Development) (2019). *Geographical Distribution of Financial Flows to Developing Countries 2019*. OECD, Paris.

Overton, J. and Murray, W. (2016). Fictive place. *Progress in Human Geography* 40:6, 794–809.

Peet, R. and Hartwick, E. (2009). *Theories of Development: Contentions, Arguments, Alternatives*, 2nd ed. Guildford Press, New York.

Potter, R., Binns, T., Elliot, J. and Smith, D. (2018). *Geographies of Development: An Introduction to Development Studies*, 3rd ed. Routledge, New York.

Power, M. (2003). *Rethinking Development Geographies*. Routledge, London.

Quadir, F. (2013). Rising donors and the new narrative of 'South – South' cooperation: What prospects for changing the landscape of development assistance programmes? *Third World Quarterly* 34:2, 321–338.

Rahnema, M. and Bawtree, V. (1997). *The Post-Development Reader*. Zed Books, London.

Rist, G. (2007). Development as a buzzword. *Development in Practice* 17:4–5, 485–491.

Rocha, H. (2013). Dominant development paradigms: A review and integration. *Journal of Markets & Morality* 16:1, 7–24.

Rostow, W. (1960). The United States in the world arena. *Naval War College Review* 13:6, 7.

Roy, A. (2010). *Poverty Capital: Microfinance and the Making of Development*. Routledge, London.

Saith, A. (2006). From universal values to millennium development goals: Lost in translation. *Development and Change* 37:6, 1167–1199.

Schulpen, L. and Gibbon, P. (2002). Private sector development: Policies, practices and problems. *World Development* 30:1, 1–15.

Sen, A. (1997a). Development thinking at the beginning of the 21st century. *Discussion Paper*. London.

Sen, A. (1997b). Editorial: Human capital and human capability. *World Development* 25:12, 1959–1961.

Severino, J. and Ray, O. (2009). The end of ODA: Death and rebirth of a global public policy. *SSRN* 1392460.

Sheppard, E. and Leitner, H. (2010). Quo vadis neoliberalism? The remaking of global capitalist governance after the Washington consensus. *Geoforum* 41:2, 185–194.

Silvey, R. (2010). Development geography: Politics and 'the state' under crisis. *Progress in Human Geography* 34:6, 828–834.

Soederberg, S. (2004). American empire and 'excluded states': The millennium challenge account and the shift to pre-emptive development. *Third World Quarterly* 25:2, 279–302.

Taylor, I. (2016). Dependency redux: Why Africa is not rising. *Review of African Political Economy* 43:147, 8–25.

Temple, J. (2010). Aid and conditionality. In Rodrik, D. and Rosenzweig, M. (eds) *Handbook of Development Economics* 5, 4415–4523. Elsevier, London.

Tomlinson, J. (2012). Cultural imperialism. In Ritzer, G. (ed) *The Wiley-Blackwell Encyclopedia of Globalization*, 366–375. Wiley-Blackwell, Oxford.

UNDP (United Nation Development Programme) (2018). *Human Development Report 2018*. hdr.undp.org/en/content/human-development-index-hdi/. Retrieved April 2019.

UNECA (United Nations Economic Commission for Africa) (2018b). *African Statistical Yearbook 2018*. UNECA, Addis Ababa.

United Nations (2000). *United Nations Millennium Development Goals*. United Nations, New York.

United Nations (2015). *The 2030 Agenda for Sustainable Development*. United Nations, New York.

United Nations (2018). *Financing for Development: Progress and Prospects 2018*. United Nations, New York.

Villanger, E. (2016). Back in business: Private sector development for poverty reduction in Norwegian aid. *Forum for Development Studies* 43:2, 333–362.

Willis, K. (2011). *Theories and Practices of Development*. Routledge, London.

Willis, K. and Kumar, M. (2009). Development I. In Kitchin, R. and Thrift, N. (eds) *International Encyclopedia of Human Geography*, Vol. 3, 111–116. Elsevier, London.

Ylönen, M. (2012). Finland, transformation of the Finnish development policy: The private turn. In The Reality of Aid International Coordinating Committee (ed) *Aid and the Private Sector: Catalysing Poverty Reduction and Development? Reality of Aid Report*, 205–210. IBON International, Quezon City.

4

INNOVATION-FOCUSSED DEVELOPMENT COOPERATION

Introduction

Investing in the systematic and interactive development of innovations has become a common policy objective of numerous governments in Africa. Often this dynamic is initiated and stimulated by international development assistance. During the last two decades, innovations have increasingly become the central object of development cooperation. Innovation-related projects are demanded by the countries receiving aid. In general, these projects are conceived to promote employment, increase productivity and foster competitive advantages in the Global South.

Innovation-focussed development cooperation is also usually seen positively in the donor communities. In addition to development impacts in the aid-receiving countries, these innovation projects enhance the knowledge and expertise of the actors in the donor countries, and thus open up new opportunities to enter emerging innovation markets in the Global South. In addition to economic growth, innovation-focussed development cooperation aims also to create socio-economically sustainable and transformative innovation policies that are tailored for the socio-economic contexts of the Global South. These new approaches are not merely policy transfers from the more mature economies (Cassiolato & Vitorino, 2009; Arocena & Sutz, 2014; Scerri et al., 2014).

- Innovation-focussed development cooperation is increasingly important in Africa in the early 21st century.

- Innovation-focussed development cooperation refers to all policies and practices in international development aid and assistance that aim to foster systemic and interactive development, and the use of innovations to improve local socio-economic development.
- Development cooperation sometimes helps donor countries' enterprises to enter the Global South markets, which is economically useful for the donor countries as well.

In this chapter, we analyse innovation-focussed development cooperation. Following this introduction, in Section 4.2, we begin by briefly discussing the role of innovation and technology in development assistance over various decades. In Section 4.3, we define and focus on contemporary innovation-focussed development. In the remaining sections, we divide innovation-focussed development cooperation practices into three themes and elaborate on those with different case examples. In Section 4.4, we analyse how African domestic endeavours, together with donor countries, aim to establish innovation systems, strengthen innovation capacity, create regional networks and develop innovation policies in African countries. In particular, we scrutinise science, innovation and technology cooperation in Africa. This is challenging in Africa because in general, universities and other knowledge-providing organisations are rather weak (see Chapter 6). In Section 4.5, we deal with local innovation development and inclusive innovations that are among the most promising approaches in Africa (see also Chapter 8). However, these themes do not function separately from each other; cross-cutting programmes link them together. In the final Section, 4.6, we wrap up and provide conclusions.

Innovation in development assistance

The promotion of science, technology and innovation has been long incorporated into specific policies and development strategies of the countries in the most advanced economies. Often, these policies have been influenced by the OECD, a key organisation providing guidelines for economic growth (Lundvall & Borrás, 2005). After World War II, during the modernisation era, new science and technology policies emphasised public investments in basic science and R&D (Schot & Steinmueller, 2018). Science and technology were considered engines of economic growth. The Global North countries concentrated on boosting their industrialisation, production for and consumption by the masses. STI policies were especially supported during the arms race of the Cold War period from the 1960s until the 1980s. Technological development and innovations were incorporated as tools for neoliberal development policies from the 1980s onward. It was conceived that countries' capacity to develop new technologies would promote their private sectors, and eventually, their overall economic development. STI policies have also been important in development assistance since the emergence of the modern ODA (see Chapter 3). Most often, the Global North donor countries offered technical assistance and transferred their innovations, such as agricultural machinery, to the Global South (Wilson, 2007). The positive results of such simple technology transfers were often short-term. Until the end of the 1990s, in the development assistance field, it was considered that innovations and new technologies would be incremental results from investments into R&D. Therefore, the development impact from successful results was to be achieved easily. Development assistance would require only straightforward transfer of technology and innovation policies from one geographical location to another.

However, later in the 2000s, the contemporary understanding of knowledge creation challenged this idea of knowledge transfer leading to innovations. Instead, knowledge creation was seen as a context-specific, interactive and complex process in which knowledge cannot be moved to other geographical contexts without changing (Hautala & Jauhiainen, 2014). That impedes the possibilities to repeat development successes based on direct innovation policy transfer from the Global North to the Global South. Therefore, systematic, interactive and contextualised knowledge creation processes are needed in fostering the emergence of innovations in the Global South (see Chapter 2).

Poverty alleviation has always been a key target of development assistance. However, in the programs and projects aiming to reduce poverty, little attention was traditionally paid to innovations and technology. Most goals concentrated on community development and enhancing basic public services, such as health care and education sectors. However, the beyond aid approach developed in the early 21st century focuses mostly on economic growth and mutual economic interests between donor and recipient countries (see Box 4.1). This has made technology and innovations important in development cooperation (see Chapter 3).

BOX 4.1   INNOVATION-FOCUSSED DEVELOPMENT COOPERATION

Innovation-focussed development cooperation refers to all policies and practices in international development aid and assistance that aim to foster systemic and interactive development and use of innovations to improve local socio-economic development. It aims to strengthen national and macro-regional innovation policy creation and networking in the Global South. It focuses on enhancing the STI capacity of aid-receiving countries and enhances local communities' capacity to utilise inclusive innovations (see Chapter 8).

Innovation-focussed development cooperation is distinct from the earlier policies leaning on STI policies and knowledge transfer. Instead, innovation-focussed development is considered to be a complex systemic process embedded into local socio-economic contexts and circumstances. With such cooperation, for the first time, development programs have started to systematically develop local countries' capabilities to develop and use innovations for their own socio-economic development.

Cooperation practices support the development and implementation of national and regional innovation policies. This means, for example, the establishment of new institutional and organisational environments. The aim is also to strengthen the capacities and networking of key actors, such as line ministries and high-level officials. Innovation-focussed development coordination is connected to the challenging endeavour of broader transformative innovation policies. It involves different stakeholders and actors and policy implementation levels. Novel innovation policies are needed, because the past approaches and decisions constrain opportunities to enhance innovation creation (Kraemer-Mbula & Wunsch-Vincent, 2016).

Internationally, the first global events contributing to the evolution of innovation-focussed development policy were the UN World Summits of the Information Society, which were held in Geneva in 2003 and Tunis in 2005. In these summits, the aim was to find solutions to the growing digital divide between the Global North and the Global South (Hooli & Jauhiainen, 2017). The first development programmes in Africa in the late 1990s and the beginning of the 2000s focussed on technical skills, ICT equipment, knowledge and innovation for development. Initially, the idea was to transfer practices, and many times second-hand ICT machinery, from the Global North to the Global South. Later, the projects became more complex, focussing on systematic evolvement of innovation systems and creating more context-specific innovation policies. New forms of cooperation emerged to organise joint programs among the public development organisations and private companies, foundations, NGOs and local communities.

The increased popularity of innovation-focussed development derives from two aspects: First, as discussed in Chapters 1 and 2, the systemic development and application of innovations substantially influences a nation's socio-economic development. Together, innovation development and application have become the key framework for socio-economic development in the Global South, and the countries of the Global North support these achievements. Innovations would have a major impact in local communities toward rising out of poverty in Africa and transforming the African countries into middle-income economies. Due to these expectations, innovation-focussed cooperation has also become desirable for aid-receiving countries in the Global South.

Second, the economic progress of countries in the Global South and the emergence of completely new pro-poor technologies and business models recast new innovation needs and provide increasingly relevant market potential for technology companies from the Global North. In the near future, hundreds of millions of Africans will need new innovations. There will be a huge market of more than one billion young adults and youth living and born in Africa in situations where digital technologies are nearly ubiquitous. The rising education levels and absorptive capacities of growing populations, the omnipresence of digital technologies and several needs created by rapid urbanisation make massive markets for innovations, which some competitors can later up-scale to the global contexts.

These rapidly changing global dynamics have made many donor countries change their internationalisation strategies and increasingly focus on emerging and developing countries. In particular, global innovation leaders such as the United States, Germany, Finland, Sweden, the United Kingdom, Japan and Switzerland conceive of innovations not only as an effective development policy, but increasingly as an internationalisation opportunity for their domestic innovation policies (Carlson, 2006). Innovation-focussed development cooperation opens up new market possibilities for donor countries' technology companies by, for example, increasing the absorptive capacity, technological capabilities and human capital of the aid-receiving countries (Banks & Hulme, 2014; Murray & Overton, 2016). For example, increased smart phone usage, digitalisation of local services, and local development of contextual mobile applications will increase the demand to extend 5G mobile network technology, and consequently open more possibilities for innovations needed for rapid and high-volume information processing.

The most economically significant donor countries for innovation development in Africa have been the United States (USAID, 2019), the United Kingdom (DFID, 2017), Germany (BMZ, 2017) and China. For example, USAID has elevated "Catalysing innovation and partnership" as one of the five main targets of its development aid (others are: providing humanitarian assistance, empowering women and girls, promoting global health and supporting global stability) (USAID, 2019; see Section 4.4). Nevertheless, Canada, Finland and Sweden were the first movers in innovation-focussed development policy in Africa (see also Boxes 4.3, 4.5, 5.3 and 6.3).

BOX 4.2 INNOVATION-FOCUSSED DEVELOPMENT COOPERATION POLICIES AND PRACTICES OF THE NORDIC COUNTRIES

The economic success of the Nordic countries Finland and Sweden has long been based on their well-functioning innovation systems. Despite being globally small in terms of population (Finland has six million and Sweden 10 million inhabitants), both countries are famous for developing several globally relevant enterprises with successful innovation portfolios and brands such as Nokia, Rovio Entertainment (Angry Birds), Kone and Neste Oil for Finland; and Ericsson, IKEA, Spotify and Volvo for Sweden.

Finland and Sweden have been global forerunners of innovation-focussed development cooperation and practices. However, their roles in international development have been different. The development aid budget of Sweden has always been among the highest (around 1% of GDP) of DAC compared with its GDP. In comparison, the development budget of Finland has been more modest. After the reduction of development aid in the 2010s, its share fell to around 0.4% of its GDP, which is well below the UN suggestion of 0.7%.

Sweden has supported innovation processes and systems in developing countries since 2003 via its Research Cooperation Programme of the Swedish International Development Cooperation Agency (SIDA). In contemporary Swedish development policy (Government of Sweden, 2016), innovations and technology are mentioned as the first topic in contemporary development cooperation. The policy states: "Today, research, innovations and technical developments offer great opportunities for low- and middle-income countries to make faster progress than was previously the case". The policy recognises important actors for technological development such as universities, businesses, civil society and government agencies (Government of Sweden, 2016: 7). In 2015, Sweden published its position paper *Support to Innovation and Innovation Systems – Within the Framework of Swedish Research Cooperation* (Government of Sweden, 2015). The paper mentions that innovation is a dynamic, systemic and interactive process that in developing countries also includes informal sectors and local community knowledge.

The first category of support for innovation systems, processes and/or innovations is taking place through Swedish bilateral cooperation with Tanzania, Uganda, Mozambique, Rwanda and Bolivia. Most activities have focussed on sectoral innovations, especially in agriculture, health, energy and habitat. This development work has been conducted via various clusters and innovation centres, like the Morogoro Metal Works Cluster in Tanzania, the Zanzibar Seaweed Cluster and the Innovation Centre in Bolivia. SIDA has also supported the coordination of cluster development in Africa via the Pan-African Competitiveness Forum.

The second category of support has been for research on innovation systems, processes and/or innovations (Government of Sweden, 2015: 11). For example, SIDA has supported innovations via the global innovation research network Globelics (Global Network for Economics of Learning Innovation and Competence; see Box 6.5). The main objective of the network is strengthening research capacity in innovation studies and exchange of knowledge between the Global North and the Global South. In addition, SIDA supports, for example, smaller research programs in the OECD and the UNESCO, and published the African Innovation Outlook. Additionally, SIDA offers different funding schemes for innovations, like Challenge Funds, the Global Health Investment Fund, and the Global Innovation Fund.

Finland's innovation-focussed development policy has been closely linked to internationalisation of Finland's innovation policy during the 21st century. Internationalisation is a key theme in the Finnish innovation policies, and innovation systems are at the core of Finland's internationalisation (Pelkonen, 2006). Until the early 2000s, the Finnish efforts for internationalisation of innovation activities focussed mainly on Europe and the United States. Later, emerging economies, especially in Asia, became more important because of fear that a new global order will fundamentally alter the foundations of the Finnish economy (Toivonen, 2014: 134).

The first major step toward innovation-focussed development cooperation was the development policy programme in 2004. It mentioned cooperation related to knowledge-based society development as a specific sector (Ministry of Foreign Affairs, 2004). One objective was to reduce the digital divide between the Global North and the Global South. Soon the Ministry of Foreign Affairs (2005) released "Development Policy Guidelines for ICT and the Information Society", which promoted ICT for development and creating a knowledge-based society.

In the early 2000s, ICT and innovations were promoted mostly through infrastructure that was included in other development programmes. After 2004, innovation systems, ICT and STI became the main objectives of several Finnish development cooperation flagship projects. The first project was the Cooperation Framework on Innovation Systems between Finland and South Africa (COFISA) from 2006–2009 (see Box 6.2). Since COFISA, Finland has financed several large and well-known innovation-focussed development cooperation projects in Africa and Vietnam (see Box 5.3 for BEAM).

The Southern African Innovation Support Programme (SAIS 2011–2015 and SAIS II 2017–2021) supported macro-regional and national innovation system development in Botswana, Namibia, South Africa, Tanzania and Zambia, and the Southern African Development Community (SADC) Secretariat members. The SAIS is a macro-regional intermediate organisation that aims to establish sustainable knowledge-sharing networks for innovation support and partnerships; strengthen human capacity related to innovation; adapt and replicate selected best practices, projects and initiatives to support practical outcomes; and build institutional and operational elements of the innovation system on national and regional levels (SAIS, 2019).

In Tanzania, Finland ran the program TANZICT – Strengthening the Innovation Ecosystem in Tanzania (2011–2015). The main objectives of this bilateral collaboration were to strengthen the information society in Tanzania and to help the government of Tanzania to achieve its socio-economic development targets. These included capacity building of ministries, revision of Tanzania's national ICT policy and building innovation programs together with local universities and communities.

Locally, TANZICT established a living-lab network that included six innovation platforms across the country (Hooli et al., 2016). Although the policy-level changes were slow and even erratic in Tanzania, the success of the living-lab network proved that innovation is not only for highly educated specialists. Instead, a community-driven innovation can be more flexible and easier to establish. A few local, committed individuals can make a difference even with a small budget (see Box 7.3). However, the approach needs to be community-driven and cannot be controlled or driven from the top down. TANZICT was supposed to continue with Finland's Support to Tanzanian Innovation System (TANZIS) Program. Despite several attempts, by 2019 it has not yet managed to start.

Innovation-focussed development cooperation also directly funds activities that enhance donor countries' knowledge and train new experts to understand and operationalise innovation activities in the Global South. Innovation systems and markets in Africa are heterogeneous, fluid and still evolving. These have special needs, requirements and structural circumstances. At the same time, they offer alluring business and innovation opportunities. However, coping in these markets requires specific knowledge. Although innovation policies in Africa are still unfolding, conceptualising innovation to the distinct socio-economic contexts helps operational practices and provides suggestions as to how this cooperation could be fostered. Co-creational processes between various actors from the Global North and the Global South are an important framework for success. Additionally, from this perspective, developmental cooperation in international aid and assistance is necessary.

The recent discussion about transformative innovation policies (Kaplinsky, 2011; Schot & Steinmueller, 2018; see Section 2.5) and the focus on innovation-focussed development cooperation in development policy bring previously separate policy paradigms closer to each other (Table 4.1). The main argument of transformative innovation policies is that it is not enough to produce innovations to satisfy mass consumption. Instead, an innovation system should transform societies toward environmentally, socially and economically sustainable development (Diercks et al., 2019). However, debate and considerations about strategic directions and operational practices to organise transformative innovation policy are only sparking in the Global North. Transformation has always been an inherent part of innovation-focussed development cooperation, and local challenges, such as environmental protection, poverty alleviation and fostering sustainable economic

TABLE 4.1 Innovation and development policy trajectories

Innovation policy paradigms	Development policy paradigms
1. STI-Policy (1945 - 1990)	**1. Modernisation & neoliberalism (1945–2000)**
Science, technology & innovation, mass production	Industrialisation, technological assistance, free markets, private sector stimulation
2. National Innovation Policy (1990–)	**2. Poverty reduction/SDGs (1990–)**
Systematic development of innovation, triple helix, cooperation	Poverty alleviation, reducing inequality, education, health care, HIV/AIDS prevention
3. Transformative innovation policy (2015–)	**3. Beyond aid/innovation (2010–)**
Innovative structural change and sustainability	Innovative structural change and sustainability, private sector support, blended financing, reciprocity

growth, have been major sources of innovation. Hence, this novel juncture between innovation and development policies can also enhance the learning processes from the Global South to the Global North – not only the other way round, as usual.

Strengthening innovation policies and systems in Africa

Nowadays, science, technology and innovation are becoming a general notion in regional, national and transnational development strategies in Africa. In general, these strategies acknowledge that the transformative influence of innovation impacts societies and economies in multiple ways. Most importantly, innovations support new employment, increase productivity and enhance the economic competitiveness of countries. Hence, innovation policy is central for economic policies.

The emergence of innovation policies and innovation systems in African countries has been influenced by international organisations, development aid from donor countries and the African countries themselves. From a global perspective, innovation is a cross-cutting element in the UN SDGs, which are the most important contemporary development reference for the donor and recipient countries. Innovations are emphasised in SDG 9 (*Build resilient infrastructure, promote inclusive and sustainable industrialisation and foster innovation*). In addition, in SDG 17 (*Partnerships for the goals*) are innovations and technology-specific targets, including South–South cooperation to access STI (United Nations, 2015).

From a pan-African perspective, the AU Agenda 2063 underlines the importance of the STI approach as the main engine of socio-economic development in Africa (African Union, 2015). To implement this objective, the AU Heads of States and Governments adopted in 2014 the Science, Technology, and Innovation Strategy for Africa 2024 (STISA-2024). This policy aims to tackle challenges that hinder development in critical fields such as agriculture, environment, energy, infrastructure, health, security, mining and water. Six priority areas contribute towards the AU Agenda 2063. The STISA-2024 boosts the application of science, technology and innovation programs across Africa through increased institutional capacity and technical skills, promotion of economic competitiveness by enhancing innovation and entrepreneurship, protection of intellectual property and improvement of innovation and research infrastructure. Moreover, the AU established the African Observatory of Science, Technology and Innovation to support African countries in the implementation of the strategy. Furthermore, UNECA (United Nations Economic Commission for Africa) has promoted innovation development by assessing the state-of-the-art of innovation development and policies of several African countries.

There are challenges in these large global and pan-African strategies. The SDGs are general principles and, despite operational guidelines, do not offer resources to implement innovation policies in practice or obligations for member states. STISA-2024 is broad in scope, without detailed attention to the special contexts and need-based innovations in Africa. It does not, for example, discuss what innovation means or how its development is distinct in different African contexts. Furthermore, innovation policy development focusses exclusively on formal institutions, R&D and STI-mode innovations occurring in universities and research organisations. Such a perspective does not take into account the innovation potential of local communities and, for example, the informal sector. Therefore, as a general strategy, the STISA-2024 does not pay enough attention to the DUI mode of innovation creation, indigenous knowledge and inclusive innovations.

Innovation-focussed development cooperation between specific Global North donor countries and selected African countries has supported the design of innovation policies and systems and brought more focus on the importance of innovations for development (see Box 4.1). Several African countries have established their own national innovation policies and strategies during the early 21st century (see Box 4.2 for Kenya, and Box 4.3 for Namibia). Among the country-specific strategies, Rwanda, sometimes called "the Singapore of Africa", pronounced in the early 2000s that it would develop its ICT infrastructure to be a knowledge-based society. Knowledge society development has been one of the main initiatives of Rwanda's president Paul Kagame, who has charismatically and firmly led the country since 2000. Kagame (2006: 5), for example, stated in 2006:

> Our move towards an ICT and knowledge driven economy is a decision rooted in the practical realities and challenges within Rwanda, but equally it has taken into account recent trends so that Rwanda can position herself to compete in the global economy. Just as it is clear that growth in the 19th and 20th centuries was driven by networks of railways and highways, growth and development in the 21st century is being defined and driven by digital highways and ICT-led value-added services. In Africa, we have missed both the agricultural and industrial revolutions and in Rwanda we are determined to take full advantage of the digital revolution. This revolution is summed up by the fact that it no longer is of utmost importance where you are but rather what you can do – this is of great benefit to traditionally marginalized regions and geographically isolated populations. In our context, it will allow us to make use of our most important and most abundant resource – Our People.

There are major challenges in the African countries' attempts to establish innovation policies, systems and practices (see Boxes 4.3, 4.4 and 4.5). Although many countries have designed and approved development strategies for two or three decades ahead, the public sector often lacks solid long-term proactive visions on how innovations could be incorporated into their strategy and implementation plans. In addition, the appropriate policies have emerged very slowly, and in many cases, they became politicised and were left without formal approval. In almost in every African country, the public investments into R&D are very small in relation to their GDP and absolute terms (UNECA, 2018: 110). The weak capacities of the university sector create further challenges in operating an STI-based innovation policy in Africa (see Chapter 6). As regards DUI-based innovation, the national innovation strategies and systems rarely pay attention to indigenous knowledge, the informal sector or the broader participation of society in the innovation processes. When indigenous knowledge is mentioned, rarely are the potential challenges of its commercialisation considered, or how the material benefits of its wisdom can be shared in the society. Lately, several donor countries have reduced policy-level cooperation, as the political processes in Africa are often slow, complex and with unpredictable new challenges, including corruption. There has been much more emphasis on initiating development projects, rather than following their development and analysing and reporting clearly about the impacts of developmental cooperation.

**BOX 4.3   DEVELOPING AN INNOVATION
SYSTEM IN KENYA – SUCCESSFUL
POLICIES AND A VIVID PRIVATE
SECTOR**

Despite its rapidly expanding population, scarce resources and severe unemployment, Kenya (54 million inhabitants; 224,000 square kilometres) has managed to transform its economy towards a knowledge society during the early 21st century. It is, in Africa, the home of a burgeoning innovation scene and one of the ICT-based innovation leaders in the continent. Several co-creation hubs, living labs and universities' incubation centres, together with private entrepreneurs and transnational companies, have created a vivid and versatile private sector and numerous innovations in Kenya. These include M-Pesa (see Box 5.4), agricultural innovations such as low-tech agricultural information platform iCow (see Chapters 5 and 8), inclusive innovations such as free public Wi-Fi network Moja and clean energy products with a youth-powered sales network for poor people LivelyHoods.

The success of Kenya is supported by the interplay between the country's tech-savvy youth entrepreneurs and the national government's ability to reform policies and increase R&D funding. Kenya launched its first innovation policy in 2006. It was supposed to implement the Kenya Vision 2030 targets for an enhanced R&D, human resource development, strengthened science and technology infrastructure, and better and increased interaction and partnership between various actors in society. Importantly, the Ministry of Education, Science and Technology was established to enhance capacity-building and innovation. The creation of a new ministry was followed by many important institutions for innovations, such as the National Commission for Science, Technology and Innovation; National Research Fund; Kenya Education Network; Science, Technology and Innovation Act; Kenya Open Data Initiative; Micro and Small Enterprises Authority and Kenya National Innovation Agency (Ndemo, 2015). Moreover, Kenya's government expenditures in R&D compared to GDP are among the highest in Africa and the National

Research Fund receives up to 2% of the national GDP (UNECA, 2018: 110, 117). However, they are still far behind the public expenditures in the Global North.

In 2009, the renewal of innovation policy led to the establishment of the comprehensive Science, Technology and Innovation Policy and Strategy in Kenya. The new policy aimed to mainstream STI in all sectors of the Kenyan economy. Perhaps the most important policy reform was a constitutional dispensation in 2010. The new constitution improved opportunities and gave equal rights to many previously marginalised groups – including women, whose previous legal status was weaker. The new constitution acknowledges the value of indigenous knowledge in innovation and technology development.

Innovation-focussed development policy had an important role in enhancing innovation policy, university collaboration and many development practices in Kenya. Among others, a long-term partnership exists between the University of Nairobi and MIT in the United States. In addition, the Republic of Korea contributed to the establishment of innovation systems in Kenya, and the Korean Institute of Science and Technology trained Kenyan innovation experts.

BOX 4.4   NAMIBIA – A LONG WAY TOWARD INNOVATION POLICY

Namibia in southwest Africa is an illustrative case of opportunities and challenges that the creation of comprehensive innovation policy in Africa encounters (Jauhiainen & Hooli, 2017). The country became independent in 1990, when it was emancipated from the South African apartheid government. It has the second-lowest population density (fewer than three inhabitants per square kilometre) of any sovereign country in the world, as its massive surface area (824,292 km²) is populated only by 2.6 million inhabitants.

After gaining independence, Namibia with its modern constitution became one of the more stable and democratic countries in Africa. The economic growth of Namibia has been rapid, though with ups and downs, and currently it is classified as an upper-middle-income country with stable macroeconomic conditions. Mining, agriculture, tourism, marine technology and fishery are the main livelihood activities. However, severe socio-economic challenges persist, especially related to inequality, poverty, unemployment, and having many people employed in an informal economy. Namibia practices close macro-regional cooperation especially with its neighbours South Africa and Botswana.

Namibia has created and implemented many innovation-related laws, strategies, policies and institutions needed for the establishment of a proper innovation system for over two decades. In the early 21st century, Namibia approved its grand vision – Vision 2030 – according to which by 2030 Namibia should be a knowledge-based, exceedingly industrial and competitive nation with sustainable development and high quality of life (Republic of Namibia, 2004). To achieve this objective, Namibia aims to establish a well-functioning innovation system that is based on interaction between the national government, universities, private companies and the civil sector (National Planning Commission, 2017).

Some policies related to innovation systems are well grounded. For example, the National Research, Science and Technology policy was established as early as 1999. However, it took 14 years to establish the first institutions mentioned in the policy. The first coherent national innovation policy was drafted in 2011, but its formal approval and establishment into practice did not proceed. Even though Namibia was an early mover in consistent innovation-related policies, it lost its early momentum as its government failed to establish innovation policies and institutions. The main industries, especially mining, have not been properly attached to any local innovation systems, and their capacity to create jobs has remained low.

As a small nation, Namibia has only two public universities. However, their interaction with the private sector technology enterprises and society and cooperation amongst each other have not been without challenges (see also Box 6.4). Namibia has 10 ethnic groups that speak nine different languages and possess unique indigenous knowledge. Namibia would have comparative advantages from the doing-using-interacting mode of innovation and utilisation of rich indigenous knowledge in the innovation system. However, the main strategic focus of the government has been the STI mode of learning, requiring analytical knowledge, human capital and a well-established innovation system.

Namibia has had innovation-focussed development cooperation projects with the international donor community. For example, Germany supported the establishment of the Namibia Business Innovation Institute, and Finland has run the Southern Africa Innovation Support Programme (SAIS), a macro-regional innovation initiative in southern Africa, since 2012.

Science, technology and innovation cooperation

Enhancing high-end science, technology and innovation capacity and related facilities in Africa has become one of the main forms of innovation-focussed development cooperation. Several large-scale STI projects have been funded, particularly by non-DAC donors. Often these projects are expensive and connected to the new beyond aid agenda of development cooperation, and decoupled from other mutual interest between donor and recipient. China and India, at the forefront, fund the establishment and building of universities and research centre facilities for numerous countries in Africa. The cooperation becomes a mixture between technical expertise, scientific collaboration and commercial linkages.

China is today a key actor in innovation-related activities in Africa. For example, a comprehensive cooperation in science, technology and innovation has been the driving force of the China–South Africa partnership for over two decades. Since 1999, the cooperation includes joint academic research in multiple fields. For example, the Department of Science and Technology of South Africa and the Ministry of Science and Technology of China have funded more than 100 intergovernmental joint research projects. This has significantly promoted the establishment of a partnership network between the research institutes in these two countries. In transportation, information, astronomy and biology, the universities and research institutions of China and South Africa have jointly established research centres and laboratories to build platforms for in-depth cooperation. For instance, Huawei established the Huawei Training Centre in South Africa in 2008, and by 2017, had trained 20,000 ICT professionals there. It further created the first Huawei Authorised Information and Network Academy (HAINA) in South Africa in 2015. From 2015–2018, over 100 South African students finished their educational programmes in the HAINA and the Huawei ICT Academy. China's new technologies, such as resource satellites and Internet services, are also widely used in South Africa. For instance, the ground system of the China-Brazil Earth Resources Satellite 04 in South Africa was launched into operation in 2015. It receives satellite data daily and serves 13 countries, including South Africa. It has been widely used for monitoring floods, land use and other developments in Africa. The cooperation in high-tech industries grows. In the high-tech industries, such as ICT, digital technology, renewable energy and automobile manufacturing, Chinese companies have planted roots in South Africa through cooperation with local enterprises, providing new impetus to South Africa's economic and social development (see Chapter 5). Although South Africa has been the oldest and closest partner of China-Africa STI-cooperation, in recent years China has extended this cooperation to almost every corner of the continent. Furthermore, the majority of Africa's 4G Internet network has been built by the Chinese company Huawei (Mackinnon, 2019).

The USAID Centre for Innovation and Impact (CII) has a clear business orientation in its approaches to development. It focuses on open innovation, innovator support, adoption of cutting-edge technologies and amplification of the use of innovation. It accelerates impact against the world's most important health challenges in Africa. For this, it invests seed money for most promising innovations and applies market-oriented approaches to transform research into new health innovations in Africa and promote their scaling-up to new markets. It has also several flagship partnerships with the private sector expertise in health issues in Ghana, Malawi, Sierra Leone and Zambia (USAID, 2019).

The AFRICA-ai-JAPAN project is a joint initiative involving Japan International Cooperation Agency (JICA), Pan African University's Institute for Basic Sciences, Technology and Innovation (PAUSTI) and Jomo Kenyatta University of Agriculture and Technology (JKUAT). This project, started in 2014, strengthens the knowledge and skills – mainly in agriculture, engineering, science and biotechnology – of both PAUSTI and JKUAT students and staff. The project is unique since it promotes the full utilisation of local and indigenous knowledge, resources, experiences and wisdom generated and accumulated in Africa to solve Africa's problems. The project promotes African indigenous knowledge and wisdom geared towards African innovation, strengthens the knowledge and ability of PAUSTI and JKUAT students and industries to actualise their innovative ideas and encourages vitality in the actions of private and public industries in Africa (Africa-ai-Japan, 2019).

France has tens of projects, mainly in Francophone Africa. Digitalisation and innovation are the main objectives of the French development organisation Agence Française de Développement (AFD). Accompanying the digital revolution, the ADF aims to enhance technologies to promote social, economic and environmental sustainable development with issues around digital transition. This includes, for example, net neutrality, cyber-security, open data, personal data protection, environmental protection and cultural diversity. Digital technology and innovations are seen as assets to accomplish the SDGs. This includes fostering the e-government and smart-city/village concepts. For example, the OPAL-project in Senegal intends to utilise private companies' data for the public good. In Côte d'Ivoire, ADF supports the GeoPoppy initiative that collects data on forested and cultivated areas. In Burkina Faso, the MobiSan project provides medical assistance via mobile phones to remote areas. The AFD Digital Challenge is a global innovation competition that encourages young start-ups' projects favouring African development. The Reducing Digital Divide programme supports large-scale infrastructure projects such as building high-speed Internet networks in Africa. Furthermore, the Developing Cultural Industries project supports cultural and creative industries, particularly in the Francophone Africa, because it has also mission of supporting the use of French language (AFD, 2019).

An example of a smaller donor country STI development cooperation is BioInnovate Africa (2016–2021), which is supported by SIDA (Bioinnovate, 2019). The strategy includes developing a knowledge-based bioeconomy in eastern Africa. It supports scientists and innovators in the region to link biology-related research ideas and technologies to business and the market. This is based on collaboration at national and regional levels and between researchers and private-sector partners. This is seen as the most secure way to translate scientific outputs into usable and commercially scalable innovations, products and technologies. The programme is based at the International Centre of Insect Physiology and Ecology (Icipe) in Nairobi, Kenya. In 2019, the BioInnovate Africa partner countries were Burundi, Ethiopia, Kenya, Rwanda, Tanzania and Uganda (see Chapter 8).

Local innovation development and inclusive innovations

Finally, many innovation-focussed development cooperation projects have a strong focus indirectly or directly on local innovations. These include boosting local tech start-up scenes, smaller grassroots projects and bottom-up approaches. Although the support to local innovation development is more heterogeneous than the policy-level and STI approaches, these projects can also be divided into two types of development activities. One the one hand, local innovations and entrepreneurship are supported in local enterprises and public sectors. On the other hand, the capacity of innovation support organisations and entrepreneurs is enhanced to develop new or improved products, processes and services with and for socially and economically excluded communities

There are thus activities that foster local innovations and entrepreneurship in local enterprises and public sectors. The donors, for example, support co-creational hubs, especially in the capital cities of African countries, on the incubation of start-up companies, the development of shared facilities, promotion of entrepreneurial culture and generation of STI innovations (see Chapters 6, 7 and 8). Different innovation funds, pitching competitions and capacity-building support for enterprises in their early stages have become very common developmental instruments.

Innovation-focussed development cooperation has brought new phenomena to the international development scene as donor countries increasingly support their own companies. Several new development programmes fund the Global North's innovative companies that aim to operate in the Global South (see Chapter 3). The support is to, for example, nurture new investments, contribute to self-regulating markets and market efficiencies, improve individual income levels by creating new and better jobs, or generate new tax revenues that governments in the Global South countries can use for social welfare and poverty reduction. For example, Finland established a new programme called Business with Impact (BEAM) to create new, sustainable businesses in developing countries. The program accelerates private companies and other actors from Finland in generating innovations for tackling global development challenges. With these innovations, sustainable and successful business are developed in both developing countries and Finland (see Box 5.3).

Furthermore, an increasing number of funds support transformative or responsible innovations. A responsible innovation is a new or considerably improved product, business model or service whose application alleviates or solves various social, economic, and environmental challenges (Bos-Brouwers, 2010). For example, the Human Development Innovation Fund (HDIF), funded by the United Kingdom, aims to identify and support innovations in Tanzania that have potential to create social impacts in education, water, health, sanitation and hygiene (WASH) across Tanzania. With a focus on market-driven solutions, HDIF catalyses the development, testing and scaling of innovative models of service delivery, information and communication technologies for development (ICT4D), and product solutions in health, education and WASH (see Chapter 8).

Also, the Danish development cooperation institution DANIDA states in its Denmark-Ghana Partnership Policy 2014–2018 that the promotion of inclusive and green growth is important in Ghana. The programme takes into account an innovative approach to synergies when greening Ghana's economy. DANIDA supported the establishment of the Ghana Climate Innovation Centre (GCIC) as an innovation hub in Ghana. The GCIC works to establish and develop local institutional capacity to support Ghanaian entrepreneurs. It develops solutions for climate change mitigation and its locally appropriate adaptations. It works with businesses in the fields of renewable energy, energy efficiency, clean transport, water supply management/purification, sustainable agriculture and domestic waste management (Ministry of Foreign Affairs of Denmark, 2018).

There are also inclusive innovation processes that aim to improve the capacity of innovation support organisations and entrepreneurs to develop new and/or improved products, processes and services with and for socially and economically excluded communities. These innovations are not typically novel or relevant at larger scale, but are new and meaningful to local contexts. Many base capacity-building on communities' most disenfranchised members, such as school dropouts, youth with multiple challenges and single mothers with poor education levels. To support these innovation activities, donors establish and fund various kinds of living labs that are a very popular mode of co-creation hubs in Africa (see Section 8.5). As a platform for local communities, a living lab creates mostly DUI innovations that respond to acute everyday challenges and provide need-based solutions for them. Contextualised living labs, such as the Afrilab network, the RLab network and several individual labs, are among the more successful organisational innovations in Africa. Hence, living labs have a remarkable impact in creating transformative change in local communities.

Another important inclusive innovation approach is to consider indigenous knowledge as a source for innovations and an asset in the innovation system (Head & Atchison, 2015). Indigenous knowledge refers to knowledge accumulated over time and unique to a given society or culture. Innovation systems supported by indigenous knowledge can potentially enhance the comparative advantages of innovation systems and make innovation policy more inclusive. The role of indigenous knowledge has been especially emphasised in the India-Africa STI cooperation, which focuses on agriculture and farming; in the AFRICA-ai-JAPAN project, with its focus on agriculture, engineering, science and biotechnology; and in many national innovation strategies, such as those of Namibia and Kenya (Jauhiainen & Hooli, 2017). Indigenous knowledge will be further discussed in Section 8.3 and Box 8.2.

**BOX 4.5 TRANSFORMATIVE INNOVATION
POLICY CONSORTIUM**

The Transformative Innovation Policy Consortium (TIPC) is a programme in 2017–2022 that combines empirical research with policy experimentation, skills development, training, communication and evaluation. It gathers policy-makers, funding agencies and scholars to collaborate for a novel STI-policy framework that contributes to tackling global societal challenges as stated in the SDGs, including inequality, climate change, employment, sustainable economic growth and development. The programme seeks up-scalable and transdisciplinary approaches, new frameworks and the harnessing of mutual learning and interaction between countries in the Global North and South. TIPC aims to shape and deliver new transdisciplinary knowledge to increase interaction between different actors of innovation, including local communities (TIPC, 2019).

TIPC is underpinned by a recent publication by Schot and Steinmueller (2018) about transformative innovation policy (see Section 2.5). It is coordinated by the Science Policy Research Unit from the University of Sussex in the United Kingdom. Its current members are innovation ministries and funding agencies from South Africa, Mexico, Sweden, Finland, Norway and Colombia. It also has several additional global association programmes that include Senegal, Ghana and Kenya.

Conclusions

Innovation has come to stay as a central objective of international development assistance. At the end of the 2010s, this is labelled "innovation-focussed development cooperation". This concept covers enhancing innovation policies in the Global South, supporting the systemic organisation of innovation development through institutions and innovation systems as well as enterprises, individuals and communities that come out with various kinds of innovations. These innovations should bring local and national social-economic prosperity, empowerment and sustainability, and in the end, lift the whole continent from underdevelopment. In the end, the goal is to narrow the developmental gap between the Global North and Africa, and therefore also end the need for contemporary development cooperation.

Innovation-focussed development cooperation is an important mechanism to create reciprocal relations between donor and recipient countries. So far, the most significant factor keeping these partners together is economic interest and the potentiality of win-win situations when supply and demand of innovations meet appropriately in the African contexts. In such cooperation, an important mutual learning process has been to figure out what innovation means in the context of Africa, what innovation systems could consist of in Africa and what kinds of local innovation policies are beneficial in Africa. These issues include pondering what actors, knowledge and institutions should be included and which should be excluded.

It is, however, evident that direct policy transfers from the Global North to the Global South fail in their goals of repeating success. In cooperation, the partners should be in equal positions, even though the process is about donors and recipients of development aid and assistance. The donors' distance from varied African contexts mean that contextual bottom-up or DUI modes of innovation have rarely been on the development cooperation agenda. Instead, more support has been given to aiming to route STI to African countries. A main focus on the STI mode of innovation will contribute to the agglomeration of economy and knowledge benefits arising from innovation processes, as is the case in the Global North (Lee, 2016). In Africa, this would mean that development incentives and funding mechanisms would be available only to a very small number of highly educated people in the national capitals of the recipient countries. In the development cooperation field, there is a need to broaden the scope to inclusive innovations (see Chapter 8) that can become key tools for transformative innovation systems eradicating poverty, increasing equality and promoting welfare among all Africans.

- Innovation-focussed development cooperation is important in Africa, but cooperation is not always equal among donors and recipients.
- Innovation-focussed development cooperation in Africa especially supports an institutional framework of innovations such as national strategies and legislation related to innovations, innovation systems and innovation policies and often STI-based activities.
- The broadening of development cooperation to include innovation supports transformative innovation systems for eradicating poverty, increasing equality and promoting welfare among all Africans.

Discussion questions

- What is meant by innovation-focussed development cooperation?
- What kind of innovation-focussed development cooperation can simultaneously support both STI and DUI modes of knowledge creation and innovation development?

- Discuss the opportunities and challenges of innovation-focussed development cooperation to solve the grand challenges in Africa such as poverty and inequality.

References

AFD (Agence Française de Dévelopment) (2019). *Towards a World in Common*. www.afd.fr/. Retrieved June 2019.

Africa-ai-Japan (2019). *African Union African Innovation JKUAT and PAUSTI Network Project*. www.jkuat.ac.ke/. Retrieved June 2019.

African Union (2015). *Agenda 2063: The Africa We Want*. www.au.int/. Retrieved June 2019.

Arocena, R. and Sutz, J. (2014). Innovation and democratisation of knowledge as a contribution to inclusive development. In Dutrénit, G. and Sutz, J. (eds) *National Innovation Systems, Social Inclusion and Development. The Latin American Experience*, 15–33. Edward Elgar, Cheltenham.

Banks, N. and Hulme, D. (2014). New development alternatives or business as usual with a new face? The transformative potential of new actors and alliances in development. *Third World Quarterly* 35:1, 181–195.

Bioinnovate Africa (2019). *About Us*. bioinnovative- africa.org/about-us/. Retrieved June 2019.

BMZ (The Federal Ministry for Economic Cooperation and Development) (2017). *Harnessing the Digital Revolution for Sustainable Development. The Digital Agenda of the BMZ*. www.bmz.de/. Retrieved June 2019.

Bos-Brouwers, H. (2010). Corporate sustainability and innovation in SMEs: Evidence of themes and activities in practice. *Business Strategy and the Environment* 19, 417–435.

Carlson, B. (2006). Internationalization of innovation systems: A survey from the literature. *Research Policy* 35:1, 56–67.

Cassiolato, J. and Vitorino, V. (2009). *BRICS and Development Alternatives: Innovation Systems and Policies*. Anthem Press, London.

DFID (Department for International Development) (2017). *Economic Development Strategy: Prosperity, Poverty and Meeting Global Challenges*. DFID, London.

Diercks, G., Larsen, H. and Steward, F. (2019). Transformative innovation policy: Addressing variety in an emerging policy paradigm. *Research Policy* 48:4, 880–894.

Government of Sweden (2015). *Support to Innovation and Innovation Systems – Within the Framework of Swedish Research Cooperation*. Government of Sweden, Stockholm.

Government of Sweden (2016). Policy framework for Swedish development cooperation and humanitarian assistance. *Government Communication* 2016/17:60.

Hautala, J. and Jauhiainen, J. (2014). Spatio-temporal aspects of knowledge creation. *Research Policy* 43, 655–668.

Head, L. and Atchison, J. (2015). Entangled invasive lives: Indigenous invasive plant management in northern Australia. *Geografiska Annaler: Series B, Human Geography* 97:2, 169–182.

Hooli, L. and Jauhiainen, J. (2017). Development aid 2.0 – towards innovation-centric development so-operation: The case of Finland in southern Africa. In Cunningham,

P. and Cunningham, M. (eds) *IST-Africa 2017 Conference Proceedings*, 1–9. IIMC International Information Management Corporation, Windhoek, Namibia.

Hooli, L., Jauhiainen, J. and Lähde, K. (2016). Living labs and knowledge creation in developing countries: Living labs as a tool for socio-economic resilience in Tanzania. *African Journal of Science, Technology, Innovation and Development* 8:1, 61–70.

Jauhiainen, J. and Hooli, L. (2017). Indigenous knowledge and developing countries' innovation systems. The case of Namibia. *International Journal of Innovation Studies* 1:1, 89–106.

Kagame, P. (2006). *The NICI-2010 Plan: An Integrated ICT-Led Socio-Economic Development Plan for Rwanda 2006–2010.* Government of Rwanda, Kigali.

Kaplinsky, R. (2011). Schumacher meets Schumpeter: Appropriate technology below the radar. *Research Policy* 40:2, 193–203.

Kraemer-Mbula, E. and Wunsch-Vincent, S. (eds) (2016). *The Informal Economy in Developing Nations.* Cambridge University Press, Cambridge.

Lee, N. (2016). Growth with inequality? The local consequences of innovation and creativity. In Sheamur, R., Carrincazeaux, C. and Doloreux, D. (eds) *Handbook on the Geographies of Innovation*, 419–431. Edward Elgar, Cheltenham.

Lundvall, B. and Borrás, S. (2005). Science, technology and innovation policy. In Nelson, R., Mowery, D. and Fagerberg, J. (eds) *The Oxford Handbook of Innovation*, 599–631. Oxford University Press, Oxford.

Mackinnon, A. (2019). For Africa, Chinese-built internet is better than no internet at all. *Foreign Affairs*, March 19.

Ministry of Foreign Affairs (2004). *Development Policy.* Ministry of Foreign Affairs, Helsinki.

Ministry of Foreign Affairs (2005). *Development Policy Guidelines for ICT and the Information Society.* Ministry of Foreign Affairs, Helsinki.

Ministry of Foreign Affairs of Denmark (2018). *Denmark – Ghana Partnership Policy 2014–2018.* DANIDA International Development Cooperation, Ministry of Foreign Affairs of Denmark, Copenhagen.

Murray, W. and Overton, J. (2016). Retroliberalism and the new aid regime of the 2010s. *Progress in Development Studies* 16:3, 244–260.

National Planning Commission (2017). *Namibia's 5th National Development Plan.* Windhoek, Namibia.

Ndemo, B. (2015). Effective innovation policies for development. In Wunsch-Vincent, S., Lanvin, B. and Dutta, S. (eds) *The Global Innovation Index 2015: Effective Innovation Policies for Development*, 131–138. Johnson Cornell University, Ithaca, NY.

Pelkonen, A. (2006). The problem of integrated policy: Analyzing the governing role of the Science and Technology Policy Council of Finland. *Science and Public Policy* 33:9, 669–680.

Republic of Namibia (2004). *Namibian Vision 2030, Namibian Framework for Long-Term National Development.* Office of the President, Windhoek, Namibia.

SAIS (Southern Africa Innovation Support) (2019). *Southern Africa Innovation Support.* www.saisprogramme.org/. Retrieved June 2019.

Scerri, M., Soares, M. and Maharajh, R. (2014). The co-evolution of innovation and inequality. In Soares, M., Scerri, M. and Maharajh, R. (eds) *Inequality and*

Development Challenges. BRICS – National Systems of Innovation, 1–19. Routledge, London.

Schot, J. and Steinmueller, W. (2018). Three frames for innovation policy: R&D, systems of innovation and transformative change. *Research Policy* 47:9, 1554–1567.

TIPC (Transformative Innovation Policy Consortium) (2019). *Transformative Innovation Policy Consortium*. www.tipconsortium.net/. Retrieved June 2019.

Toivonen, H. (2014). Knowledge economy and globalization. In Halme, K., Lindy, I., Piirainen, K., Salminen, V. and White, J. (eds) *Finland as a Knowledge Economy 2.0. Lessons on Policies and Governance*. World Bank, Washington, DC.

UNECA (United Nations Economic Commission for Africa) (2018). *Africa Sustainable Development Report 2018. Towards a Transformed and Resilient Continent*. UNECA, Addis Ababa.

United Nations (2015). *The 2030 Agenda for Sustainable Development*. United Nations, New York.

USAID (2019). *Catalyzing Innovation and Partnership*. www.usaid.gov/catalyzing-innovation-and-partnership/. Retrieved June 2019.

Wilson, G. (2007). Knowledge, innovation and re-inventing technical assistance for development. *Progress in Development Studies* 7:3, 183–199.

5

PRIVATE SECTOR

Introduction

Sustainable economic growth with innovations is currently the main objective of international development cooperation (see Chapter 4). The private sector is considered the main engine of development and economic growth in the Global South and a key actor in its well-functioning innovation system (Mawdsley, 2017). The private sector is expected to create new and better jobs, nurture new investments, promote market efficiency and self-regulation, increase tax revenues and develop need-based innovations in the Global South. These actions improve social welfare, raise individuals' income levels and alleviate poverty (Jeppesen, 2005; Di Bella et al., 2013).

The global boom of innovations, start-up enterprises and technology entrepreneurship has gained a foothold in Africa. The local innovative private sector is expected to drive the continent's social transformation and economic growth (Friederici, 2018). Notwithstanding general claims about inclusive growth and economic trickle-down development, it is very difficult to measure and prove the impact of development cooperation led by the private sector. Hence, more careful analyses, new policy tools and operational practices are needed to better connect the private sector's involvement in general development objectives.

In this chapter, we discuss the private sector's manifold but complex role in international innovation-focussed development cooperation and the private sector's innovations for development in Africa. After this introduction, we explain in Section 5.2 the private sector's role in development cooperation. International development cooperation and funds have supported several technology initiatives in Africa. In Section 5.3, we analyse the recent innovation boom in Africa and focus on technology development and frugal innovations. We illustrate how the mobile phone revolution and rising local start-up entrepreneurs are expected to be the engine of economic growth in Africa. Finally, we present in Section 5.4 our conclusions and reflect on the kind of development the private sector's increased role in development cooperation provides and for whom it is provided.

- Contemporary international development cooperation promotes businesses and related institutional environments via the private sector in the Global South, including Africa.
- The rising and up-scaling of local African technology start-ups, with their frugal innovations, are hoped to unlock the continent's economic growth, so technology entrepreneurship has become a main tool in international development cooperation.
- The support for innovative entrepreneurs brings new actors, knowledge and finance to development cooperation; however, it may also increase inequalities in Africa.

The private sector in development cooperation

The private sector has many roles in the development of Africa (see Box 5.1). On the one hand, some enterprises foster immanent development by simply operating in Africa. On the other hand, some private enterprises persistently act toward positive development in the continent and in its countries and regions (Blowfield & Dolan, 2014: 23–26). These two roles of the private sector in development are closely interconnected (McEwan et al., 2017).

BOX 5.1 PRIVATE SECTOR

The private sector is run by private individuals and enterprises for profit. Its role is important in a knowledge-based society because it develops and commercialises innovations, and it often invests in the R&D behind many innovations. It interacts with other key knowledge producers and exploiters, such as universities, product and service innovation end-users and governments supporting innovation policies and systems. The private sector includes a heterogeneous group of actors with various motivations and backgrounds in development cooperation. Private companies vary, from large transnational corporations and technology giants to small family businesses and start-ups to individuals in the informal sector.

During the early 21st century, private enterprises have become significant actors in international development. Numerous development agencies, such as the UNDP, recognise private companies' role as central drivers of economic progress in the Global South. They are an important source of innovations, new technologies and socio-economic investments in development (UNDP, 2012).

The private sector has various roles in official development cooperation. Several donors believe their domestic companies conducting normal business activities in the Global South will increase investments, create innovations and enhance employment. Therefore, the development cooperation between the countries donating aid and the countries receiving aid is increasingly based on mutual economic interests. Contemporary development cooperation promotes businesses and related institutional environments via the Global South's private sector.

The private sector has always existed in international development cooperation. During the early 21st century, however, its role in development has intensified and spread to many fields (Eurodad, 2013). Scholars who analysed the increasing importance of the private companies in the development led by economic growth have noted such changes in the Global South, including in Africa (Tomlinson, 2012b; Di Bella et al., 2013; Blowfield & Dolan, 2014; Mawdsley, 2015; McEwan et al., 2017).

Several international agreements supported the private sector's rise in development cooperation. In the early 21st century, several international high-level fora on aid effectiveness were held to modernise, deepen and broaden development cooperation and aid delivery. One particular event to intensify development cooperation led by the private sector was the 2011 forum held in Busan, South Korea. Thousands of delegates from over 100 countries discussed novel frameworks for development cooperation. The declaration of the Busan Partnership for Effective Development Co-operation agreed that development requires active participation from various public, private and non-governmental stakeholders. Notably, the private sector was, for the first time, officially acknowledged as a key actor in the strategic design and practical implementation of international development policies and practices (Eyben & Savage, 2013).

In continuation of this, the private sector representatives had an active role in preparing and launching the UN Agenda 2030 SDGs. As discussed in Chapter 3, by the early 2020s, SDGs became the most important targets for global development. Therefore, most donors' current development objectives are based on SDGs. The private sector is considered an active central stakeholder and agent in development dynamics, rather than a philanthropic sponsor (Scheyvens et al., 2016). Nowadays, key global development agencies continuously develop novel priorities, programmes, institutional structures, instruments, practices and public-private partnerships to leverage, support and finance private sector activities in development cooperation.

In general, the private sector can substantially support development. Di Bella et al. (2013: 2) distinguished three approaches inside private sector development in the Global South, including Africa. The first approach is when development organisations, donor countries and financial institutions promote private enterprises in the Global South. This includes support for local private sector actors, relevant institutions and policy development to foster the local business environment and the promotion of donor countries' enterprises and investments in countries receiving aid. The practices to help the private sector be a development actor in the Global South are important.

The second approach is how the private sector supports development. It refers to private enterprises' activities in their core businesses and operations that deliberately promote positive developmental impacts and enhance economic progress in the Global South (Di Bella et al., 2013: 2). This approach is closely connected to the discourse of corporate responsibility, which has become an important topic for credibility and image of international enterprises. Corporate responsibility is often defined as an integration of private companies' self-regulation into their business strategy to take responsible actions concerning environmental, economic and social issues (Rasche et al., 2017). In the Global South, private companies' labour practices, environmental protection and transparent taxation must be acknowledged with extra attention (Rowden, 2011; Dhahri & Omri, 2018). However, many Global North countries' development policies are seen to be enough if their domestic companies create business activities in the Global South and follow local laws, norms and acceptable business ethics. The enterprises do not necessarily need to broadly consider what their innovations mean for development in, for example, Africa.

The third approach to the private sector's role in development is when it actively seeks positive development influences in the Global South beyond core business actions (Di Bella et al., 2013: 2). These include, for example, inclusive value-added chains or business models, coordination or support for development projects and activities and active maintenance of strategic and operational objectives for responsible businesses. The enterprises consider their innovations as tools for broader development, in this case for Africa. This approach refers also to inclusive innovation as a process and a performance outcome that benefits the disenfranchised (see Chapter 8).

The private sector's involvement is bringing new actors, knowledge and much-needed resources to international development aid. The well-functioning private sector also drives economic growth and new employment. Innovation funds and accelerator programs supporting local start-up entrepreneurs are one major trend in Africa's innovation-focussed development cooperation (Box 5.2). The donors especially support promising start-ups that harness innovations to tackle challenges regarding social, economic and environmental sustainability. Programs also exist for specific interest groups, such as young entrepreneurs, women in business and SMEs. In addition, Africa is mushrooming with app competitions and awards for successful entrepreneurs and innovators, such as the annual Innovation Prize for Africa, the African Innovation Competition, the Innovation Prize for Africa, the Africa Energy Innovation Competition and the Africa Innovation Challenge. Some of these are supported by the Global North donors.

BOX 5.2 DEVELOPMENT COOPERATION SUPPORTING TECHNOLOGY DEVELOPMENT FOR AFRICA

The USAID and the United States Department of State are major donors of international aid to Africa (see Chapter 4). In 2015, they provided more than $8 billion USD to 47 countries in Sub-Saharan Africa. The USAID's US Global Development Lab is important for innovation development. It is targeted to several regions in the world, but Africa and Asia are its main areas. One of its activities is a year-round grant competition (Development Innovation Ventures) for innovative ideas, pilots and testing. Another modality is scaling up solutions that demonstrate widespread impact and cost-effectiveness. For example, the Scaling Off-Grid

Electric initiative aims to connect 20 million households in southern Africa that are currently without access to a grid. The aim is to provide them with solar energy by 2030 by supporting solar companies in the region. Similarly, the Digital Inclusion initiative aims to bridge the digital divide and expand low-cost access to the Internet in Southern Africa.

The Human Development Innovation Fund (HDIF) provides grants to businesses, research institutions and NGOs for scaling innovations with a focus on market-driven solutions to create a social impact. Its partners are the United Kingdom's Department of International Development (DFID), the Tanzania Commission for Science and Technology (COSTECH) and the KPMG in East Africa. Its activities annually reach over £40 million GBP. The HDIF catalyses the development, testing and scaling of innovative models for service delivery, for development-focussed ICTs and for product solutions in education, water, health, sanitation and hygiene. It has managed tens of development projects in Tanzania.

In technology, HDIF has supported, for example, the development of Nanofilter, which is a local innovation for a low-cost water filtration and purification system. Nanofilter has received tens of awards, employed hundreds of people and impacted over 250,000 people who can now reach clean and safe water. Another example is CliniPAK (clinical patient administration kit), a portable and wireless data collection and reporting kit used in Zanzibar, Tanzania. It is suitable for people with limited computer literacy in areas without reliable electricity. The innovation comes from Global Health Initiative, a jointly run organisation between Verna Cares, the charitable trust of the United States technology and robotics company Verna Technologies in Cambridge, MA, and the United States' Office of the National Coordinator for Health Information Technology.

Several authors in development studies and various actors in NGOs have been concerned about private enterprises' ambiguous role in development assistance, despite many of these activities supporting the lives of the people in the Global South (Schulpen & Gibbon, 2002; Kolk & van Tulder, 2006; Davis, 2012; Tomlinson, 2012b; Kindornay & Reilly-King, 2013; Blowfield & Dolan, 2014; Mawdsley, 2015; Hooli & Jauhiainen, 2017; Mawdsley, 2017; McEwan et al., 2017). However, Blowfield & Dolan (2014: 24) directly state that a private company "is no more responsible for development outcomes than a hammer is responsible for the carpenter's thump".

One challenge concerning private companies acting for development in Africa is their broader accountability concerning development. The donors rarely consider deeply how the development assistance objectives and the profit-focussed business activities match (Tomlinson, 2012a; Kindornay & Reilly-King, 2013; Mawdsley, 2015). The support to the private sector, the focus on economic development and business interests may reduce donors' focus on poverty and inequality alleviation in development cooperation and aid. The private sector's activities in innovation development require skilled individuals and advanced knowledge. The poor people who are the primary targets of development assistance generally lack these skills and knowledge. Technology development often requires a long-term commitment and R&D and short-term risks. The disfranchised communities in poverty can seldom provide these requirements, so technology-driven development may actually expand inequalities (Cozzens & Kaplinsky, 2009). Moreover, when underdevelopment and poverty are regarded as potential business opportunities, the development cooperation targets may focus merely on entrepreneurial and competitive individuals (Hooli, 2016: 69). Since the early 2000s, for example, Finland has transformed its development policy for the Global South to engage more private sector activities and entrepreneurial individuals. In the 2010s, a specific programme was launched to create businesses with development impact (Box 5.3).

BOX 5.3   BEAM – BUSINESS WITH DEVELOPMENT IMPACTS

Donors increasingly use development assistance to subsidise and support their domestic companies operating in the Global South. The BEAM (Business with Impact) programme was the first collaborative project between Business Finland (the national funding agency for innovation and technology) and the Ministry for Foreign Affairs of Finland (responsible for Finnish development assistance). The programme's overall objective, operational in 2014–2019, was to enhance new sustainable businesses in the Global South. The goal was to subsidise Finnish technology companies' responsible innovations concerning global development challenges and simultaneously create successful sustainable businesses for them in the Global South, thus having a positive economic impact in Finland as well.

Most technology companies involved in BEAM were SMEs. They developed various technological innovations, from water and solar energy to gaming and tourism. As BEAM evidences (see Hooli & Haaranen, 2019), the private sector's role in development assistance was heterogeneous and complex. For most companies, the BEAM activities were their first foray into Global South markets. Without BEAM, they would not have entered these markets. Companies were investing their own money and time in these projects, so they brought new financial resources, actors, knowledge and innovations to development cooperation.

The objectives, target groups and location of these private sector projects differed significantly from the Finnish development policy's objectives (Figure 5.1). Finland had nine cautiously chosen official bilateral development cooperation countries. However, any ODA country listed by OECD-DAC, excluding China, was eligible for BEAM activities. The geographical focus of development cooperation led by the private sector was much more scattered and focussed on middle-income countries, whereas Finland's development policy targeted the least-developed countries. Of 36 BEAM projects' target areas, eight (22%) were upper-income countries, 21 (58%) were middle-income and seven (20%) were lower-income countries.

In BEAM, the private sector's development cooperation activities had only weak links to conventional development objectives: namely, those to alleviate poverty and inequality. Their focus was on the stable and developed markets of middle-income countries' capitals, where their main partners were educated and wealthy minorities. The majority poor of local communities in developing countries were hardly involved in the project design and implementation. Their role was mainly as potential end-users and consumers of the final products and services. BEAM was a mixture of politically steering development targets and hoping for economic profits for businesses, which proved to be challenging to reach within one united programme.

Frugal innovations, the African mobile revolution and home-grown technology geeks

The African innovation scene is evolving and diversifying rapidly. Instead of trying to cover the broad spectrum of private enterprises operating in all possible fields, we focus mostly on the emerging generation of new innovation

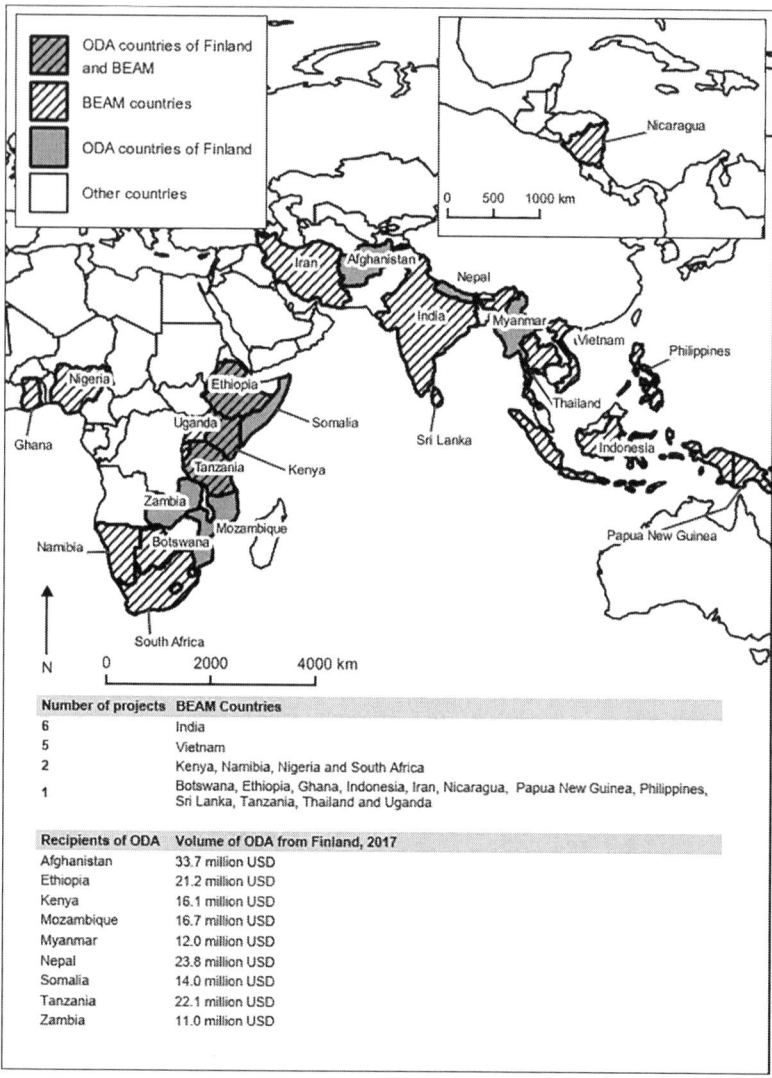

FIGURE 5.1   BEAM target countries and Finnish ODA countries

Source: Modified from Hooli & Haaranen (2019)

developers, especially from the flourishing African ICT start-up culture. These are increasingly relevant for development in Africa and for related international development cooperation.

An extensive distribution of mobile technology and rapidly improved connectivity to the Internet have enhanced access to knowledge, markets and services all over the continent. However, scholars (see Friederici et al., 2017) have made also critical remarks about technological determinism in the African ICT policies proposing self-evident positive and transformational impacts of the Internet, however, without clear evidence of its usage's impacts. Furthermore, in Africa, self-employment in the informal sector has always been more common than being employed by someone in an enterprise. This gives a particular background and flavour to the global phenomena of home-grown young entrepreneurs in Africa and how they develop new digital applications for smart phones.

In Africa, the factors behind innovations are often developers' experiences and practices, their application of existing knowledge to enhance existing goods and services and the improved design and contextualisation of products or services for various African contexts. In other words, many interesting and promising innovations in Africa are based on synthetic and symbolic knowledge and on the DUI mode of learning (Kraemer-Mbula & Wamae, 2010, see Section 2.2). This mode utilises indigenous and local knowledge, cultural interaction and an understanding of local challenges and needs, often creatively put together by private enterprises and entrepreneurs. Such innovation development interacts with local communities, uses bottom-up approaches and must be an open process connected to local socio-economic contexts.

The global poor have very rare opportunities to participate in scientific and technology-based innovation processes. Therefore, need-based innovations (also called responsible innovations) are particularly significant for their development opportunities (Halme & Korpela, 2014). These innovations are defined as novel or substantially improved products, services or business models whose application solves or alleviates local environmental, social or economic challenges (Bos-Brouwers, 2010). These kinds of innovations compensate for broader structural problems in Africa related to, for example, unreliable energy infrastructures, inadequate waste management and inefficient food production and distribution. Many African-originated innovations have frugal characteristics and have emerged to answer the local needs and to resolve local challenges.

Frugal innovations are specifically (re)designed affordable products, services, systems and business models originating from scarce resources. In frugal innovations, the complexity and costs of the total lifecycle, production and marketing are reduced (Hossain et al., 2016; Leliveld & Knorringa, 2018). The cost efficiency signifies the affordability that makes frugal innovations especially relevant for the Global South markets, where there is a high demand for functional low-cost products and services (Anggadwita et al., 2016). One example is the rising African car industry that aims to manufacture affordable cars for middle-class and poor African consumers. Such cars are manufactured, for example, in Nigeria, Ghana and Kenya. Frugal innovations specifically account for local knowledge about contexts and needs and adapt new technologies into them. For example, they have brought about several necessary agricultural and medical innovations in Africa. At the same time, however, frugal innovation may mean new cheap consumption goods for expanding markets and targeted to the growing middle-class and poor populations in Africa. Despite aiming to enhance functionality, the decreased production cost of frugal innovations means that many enterprises must make compromises in their environmental and labour standards.

A crucial factor facilitating Africa's technological development and digital revolution has been the rapid diffusion of mobile phones and expanded access to the Internet. This diffusion has been very fast, reaching even rural villages without access to fixed electrical lines. Many regard mobile phones as the most important recent and widely diffused innovation in Africa. Asongu and Nwaschukwu (2016) claim that in Sub-Saharan Africa, inclusive human development in knowledge diffusion is persistently conditional on mobile phones. The main facilitators for the rapid diffusion of mobile phones have been the inexpensive price of handsets, the improved general connectivity and the easily achievable prepaid subscriptions. During the 2010s, smart phones diffused rapidly as their price decreased. The price of a second-hand Chinese Android phone was less than $50 USD at the end of the 2010s, which is affordable for the masses in Africa. Many frugal and other innovations rely on the extensive diffusion and use of smart phones and their applications in Africa (Box 5.4).

BOX 5.4   M-PESA AS A FLAGSHIP FOR AFRICAN TECHNOLOGICAL INNOVATIONS

The most famous African mobile application is the mobile money transfer, financing and microfinancing service M-Pesa, launched in 2007 by Vodafone for the Kenyan mobile network operator Safaricom. Its development was partially funded by the British development agency DFID and was based on a Kenyan student's patented innovation. In the early 2000s, Kenyans were already using prepaid airtime to settle debts and transactions. M-Pesa then became a mobile payment method that allowed users to store value in their mobile phone's SIM card. Thus, users have a mobile account with electronic currency that allows users to quickly deposit, withdraw, transfer and convert money and to pay for goods and services (Ndung'u, 2018). The user does not need a bank account or access to a physical bank. The mobile money transfer has become very popular in Kenya and by 2013, three out of four over-15-year-olds in Kenya had a mobile account (Van Hove & Dubus, 2019). By 2018, M-Pesa had expanded to eight countries (Kenya, Tanzania, Afghanistan, South Africa, India, Mozambique, Lesotho, Egypt and Romania) and 33 million customers made over 1,000 transactions every second. However, its main success and activities are still in Kenya.

The private sector's M-Pesa transformed the lives of millions of Africans who could not use banking services before. It helped lift around 194,000 Kenyan households out of poverty between 2008 and 2015 (Suri & Jack, 2016). The use of M-Pesa has also helped to balance the individuals' needs and use of money by generating the possibility for smaller but more frequent transactions (Aker & Mbiti, 2010). Moreover, mobile money transfers helped save lives in many humanitarian crises, as it is possible to send money straight to those who need it. However, Van Hove & Dubus (2019) argue that only a minority of Kenyans use M-Pesa to save money on their mobile phone, especially due to negative interest rates and cultural practices. In addition, by 2013, the use of M-Pesa was still substantially lower among poor people, uneducated people and women. Thus, mobile savings are not a straightforward tool for financial inclusion.

M-Pesa was a forerunner of Internet banking, and it contributed to global changes in the banking sector. M-Pesa was an essential breakthrough for Kenya's technology sector and a huge inspiration for the entire technology sector in Africa. It proves that local needs can be a valuable source of novel global innovations and that it is possible to reach and implement such innovations in Africa. Safaricom pronounced its commitment to several SDG goals in its strategy and practices.

Mobile phones became common in Africa in the early 2000s (see Aker & Mbiti, 2010). The first phones were simple GSM phones with basic calling and SMS messaging functions. In 2005, the number of mobile phone subscriptions in Africa was 87 million. It grew to 366 million by 2010, and it reached 781 million in 2018 (ITU, 2019). The expansion of mobile phones has connected people in various localities to national and international information and communication. A Cameroonian fisherman, an entrepreneur in Algeria and an informal merchant from Angola now have potential access to information, markets and people beyond their immediate local community. The rapid absolute growth in the number of mobile phone subscriptions is expected to continue in the 2020s, making Africa the fastest growing large mobile communication market.

The diffusion of the Internet and the reduction of the digital divide in Africa (i.e. the access, use and impact of digital devices) are because of inexpensive handsets and changes in the infrastructure and institutions. In the early 2000s, the Internet in Africa still relied on satellites that provided slow and expensive connections. During the 2010s, broadband and underwater fibre optic cables became more common. All major cities are now connected to the 4G network, and 5G networks are being introduced in the most advanced places. In 2005, 15 million Africans were using the Internet. By 2010, Africa's number of Internet users had risen to 54 million, and there were 14 million active mobile broadband subscriptions. By 2018, the amount of active mobile broadband subscriptions had rapidly risen to 305 million (ITU, 2019). Access to the Internet is now vital in Africa for many economic, social and political activities. For example, Evans (2018) illustrates how the diffusion of the Internet and mobile phones has a significant causal positive relationship with the financial inclusion of poor Africans.

For technological innovations, smart phones and access to the Internet are obviously not enough. One also needs networks of people with relevant skills, supportive services, funding and private enterprises. One major driver of local technological development in Africa has been the rapidly emerging co-creation hubs and their networks. Hubs are popular platforms for organising innovation development activities in Africa. Hubs can be seen as a methodology to support development, an environment that supports innovation and an approach for gathering multiple stakeholders to collaborate on the promotion and development of (open) innovations (see also Chapters 2 and 6). They are platforms for development actors looking for long-term partners and platforms for short-term visitors interested in talents, trends and local innovation buzz. Hubs connect the local technology community with international entrepreneurs, donors and other actors looking for technology or business partners from Africa (Friederici, 2018).

Clustering relevant development actors into hubs benefits knowledge creation and innovation development. In Africa, these hubs are often specific to selected sectors, and do not gather developers and businesses from across sectors. The diffusion of mobile technologies, the Internet and the rapidly growing digital market have created large clusters of technological development. Today, major African cities such as Cairo in Egypt, Cape Town in South Africa, Nairobi in Kenya and Lagos in Nigeria have important technology clusters. In 2019, Cape Town had 450 to 550 high-technology enterprises, together employing 40,000 to 50,000 people. The vast majority of companies were small or medium-sized, and 3% of such companies had more than 100 employees. Almost every second enterprise was in the e-commerce or software sector, and one out of seven were in financial technologies. In addition, Africa's highest-valued technology company, Naspers, was in Cape Town. In 2018, the high-technology sector in Lagos employed 9,000 people and in Nairobi the number was 7,000 people. These clusters can be thematically specific, such as those regarding mobile technologies in Kenya and those regarding health in South Africa.

Most hubs in Africa are smaller sites for promoting and supporting technology start-ups. The most common co-creational hubs are accelerators, incubators, fab labs, co-working spaces, hackerspaces, makerspaces and other kinds of specific innovation spaces. They focus on the incubation of start-up companies, entrepreneurial culture, shared facilities and STI based innovations. They are most often located in a country's capital and other big cities (Mulas et al., 2017). By the end of the 2010s, nearly 1,000 hubs existed in Africa. The number almost doubled during recent years, and it is increasing. However, while new hubs emerge rapidly, many also disappear quickly, as hubs are rather easy to establish but challenging to sustain in rapidly developing business environments. In Africa, living labs are another type of grassroots hubs in local communities (see Chapter 8). During the last few years, as the market is maturing, hubs became increasingly specialised. Some hubs focus on incubation, and others are R&D centres. Programmes focussed on local industry are also becoming more popular in Africa. Some operate in very specific sectors, such as block chain technology solutions (BitHub Africa in Kenya) for energy and financial access in Africa.

The role of co-creational technology hubs is especially meaningful for Africa's development. They cluster and connect otherwise dispersed grassroots digital entrepreneurs, organisations, investors and mentors in the same place, they connect local actors to international actors, organise related events and are supported by open innovation processes. They are regarded as an important source for locally developed software and mobile applications (Schwartz et al., 2014). Furthermore, specialised hubs allow start-ups to differentiate, develop original services and products and find new competitive advantages. By intersecting with policymaking, research, education, social impact and business acceleration, technology hubs can break down silos that hindered innovation development in Africa. However, a typical challenge in Africa is the weak link between the hubs and academic centres (Hooli et al., 2016). Nevertheless, co-creation hubs have become important tools for international development cooperation. Several hubs have been partly or fully funded by international donors. These include Active Spaces (Cameroon), kLab (Rwanda), Buni (Tanzania), Hive Colab (Uganda), BongoHive (Zambia), Nailab and iHub (Kenya), Banta Labs (Senegal) and CcHUB (Nigeria).

Smart phones and technology start-ups have become symbols of the digital economy in cities all around Africa. The economic growth, emerging digitalisation and positive progress of several African countries have increased interest and investments in Africa. For example, venture capital investments in African technology and start-up companies expanded rapidly. They amounted to $726 million USD in 2018, which is a substantial and rapid rise from the $277 million USD in 2015. In 2018, the number of related deals increased by 127% compared with 2017, and despite two-thirds of investment taking place Nigeria, South Africa and Kenya, the remaining deals had much broader geographic distribution across Africa than in earlier years (Wee Tracker, 2018). Notwithstanding this rapid growth, the investments are still marginal. For example, in Singapore, with five million inhabitants, the venture capital investments in 2018 added up to over $1 billion USD (i.e. more than in all African countries together [Singapore, 2018]).

So far, few examples exist of systematic innovation-oriented R&D development by African companies, especially for the continent's large population. Nonetheless, some significant STI mode of learning-based innovations have emerged that foster Africa's transformation. The export of STI-based technologies is not a solution for Africa. Barasa et al. (2019) illustrate that African countries' efficiency will diminish if they invest in foreign technology in isolation from absorptive capacity-enhancing innovation inputs. Instead, many of the most influential innovations in Africa have been simple. For example, iCow is a simple mobile phone-based agricultural information platform for small farmers. By using it with SMS, no smart phone is needed; it helps the cow farmers maximise breeding potential by tracking their animals' fertility cycles. There are also tools to identify soil types, correct seeds and some crop-related problems (see Chapter 8).

Most technology-related innovations for development led by the private sector deal with people's everyday life and leisure, but for economic profit. For example, Jumia is Africa's leading e-commerce platform with an e-marketplace, and with logistics and payment services. It has generated indigenous solutions to attract offline customers and customers without a proper street address by utilising thousands of commission-based sales agents in Nigeria called J-Forces. They interact with communities, help them shop online and provide street addresses for delivery (Jumia, 2019). Business purposes can thus connect with some pro-poor solutions. Furthermore, irokotv is a very popular paid movie online movie streaming service providing and producing Nigerian Nollywood movies. Many of its customers watch these films from abroad, thus helping the Nigerian diaspora stay culturally connected to the motherland. This innovation also supports the growth of Africa's popular film industry.

The profit-oriented technology development led by the private sector also has less promising outcomes. Many successful technology innovations in Africa were developed by giant network operators, other multinational corporations and foreign enterprises. Many innovations are top-down products meant for middle-income and high-income consumers. Some innovations address the poor in rather unethical ways. For example, improved access to the Internet also created a huge online gambling market in Africa. According to a Nigerian news agency's poll, over 60 million Nigerians bet online daily, especially the youth (Anazia, 2019). On the other hand, this betting has created opportunities for a very rapid economic growth of local technology start-ups in the gaming industry, such as Paystack and Flutterwave. Moreover, social media illiteracy has generated a wide distribution of fake news, which seriously affects individuals' welfare and even the continent's democratic development. To fight against this, several donor organisations have funded independent fact-checking companies, such as Africa Check, which is sponsored by the Bill and Melinda Gates Foundation.

Africa's rapid economic growth and increase in potential have simultaneously created new desirable markets for innovation-oriented companies from the Global North (Taylor, 2016). An increasing number of northern high-technology companies search for new opportunities to refine their existing innovations and identify new needs for innovation – such as the gaming industry, to mention one dubious example discussed previously. A particular challenge arises from the unrealistic expectations and poor understanding of many private enterprises from the Global North operating in Africa. They usually have limited knowledge of local contexts, lack local cooperation partners and have too little time to be locally present in the areas in which they operate.

Another downside is wasted innovations that the private sector has produced but failed to engage with development. For example, many new affordable devices for purifying and filtering water in Africa have been developed. However, such affordable and technically solid devices do not necessarily provide commercial success, as water usage and practices are also bound to local contexts and cultures. Therefore, in the private sector's innovation for development, particular attention should be paid to home-grown African companies that are closer to local communities and their needs.

Conclusions

Contemporary international development cooperation can hardly operate without the private sector. Enterprises are needed to operationalise development goals. They are fundamental in achieving innovations and are key actors in the innovation systems needed for development in Africa. On the one hand, the private sector's involvement has brought new actors and additional funding to development cooperation. On the other hand, the private sector's activities often have weak ties to the conventional objectives of development (i.e. the reduction of poverty and inequality). In addition, donor countries increasingly allocate their development funds for their private sector enterprises operating in the Global South.

Innovations, technology start-ups and co-creational hubs are common buzzwords in the development cooperation between the Global North and Africa in the early 21st century. The development policies of donor countries hope that Africa can participate in the global digital economy through innovative local enterprises that become the engines of rapid socio-economic development and transformation. Expectations are especially high concerning the transformative pro-poor and need-based inclusive innovations (see Chapter 8). These offer solutions to various social, economic and environmental challenges in Africa and among the global poor outside of Africa.

The development impacts of technology companies differ. Some innovations are simple goods that increase consumption by growing middle- and upper-class populations. However, the need-based inclusive innovations address the local needs of the poor majority and are designed to tackle challenges caused by poverty. Supported by international development cooperation, inclusive innovations build local capacity and strengthen the local innovation system (see Chapter 8).

- Current international development cooperation and practices focus on building and strengthening the capacity of Africa's emerging local private sector.
- African technology-related start-ups possess a great opportunity for market disruption and rapid economic growth, but their future also contains potential major risks of failure.
- Innovation-driven development aid in Africa needs to be responsible.

Discussion questions

- What is the private sector's role in contemporary development cooperation?
- What opportunities and challenges do African technology-related start-ups have in responding to the grand challenges in Africa, such as poverty and inequality?
- Discuss why the private sector has become an important stakeholder in development cooperation in Africa.

References

Aker, J. and Mbiti, I. (2010). Mobile phones and economic development in Africa. *Journal of Economic Perspectives* 24:3, 207–232.

Anazia, D. (2019). Betting: Dangerous pastime for unemployed Nigerian youths. *The Guardian*, April 6.

Anggadwita, G., Ramadani, V., Alamanda, D., Ratten, V. and Hashani, M. (2016). Entrepreneurial intentions from an Islamic perspective: A study of Muslim entrepreneurs in Indonesia. *International Journal of Entrepreneurship and Small Business* 31:2, 165–179.

Asongu, S. and Nwaschukwu, J. (2016). Mobile phones in the diffusion of knowledge and persistence in inclusive human development in sub-Saharan Africa. *World Development* 86, 133–147.

Barasa, L., Vermeulen, P., Knoben, J., Kinyanjui, B. and Kimuyu, P. (2019). Innovation inputs and efficiency: Manufacturing firms in Sub-Saharan Africa. *European Journal of Innovation Management* 22:1, 59–83.

Blowfield, M. and Dolan, C. (2014). Business as a development agent: Evidence of possibility and improbability. *Third World Quarterly* 35:1, 22–42.

Bos-Brouwers, H. (2010). Corporate sustainability and innovation in SMEs: Evidence of themes and activities in practice. *Business Strategy and the Environment* 19, 417–435.

Cozzens, S. and Kaplinsky, R. (2009). Innovation, poverty and inequality: Cause, coincidence, or co-evolution? In Lundvall, B., Joseph, K., Chaminade, C. and Vang, J. (eds) *Handbook of Innovation and Developing Countries: Building Domestic Capabilities in a Global Setting*, 57–82. Edward Elgar, Cheltenham.

Davis, P. (2012). Re-thinking the role of the corporate sector in international development. *Corporate Governance* 12:4, 427–438.

Dhahri, S. and Omri, A. (2018). Entrepreneurship contribution to the three pillars of sustainable development: What does the evidence really say? *World Development* 106, 64–77.

di Bella, J., Grant, A., Kindornay, S. and Tissot, S. (2013). *The Private Sector and Development: Key Concepts*. North-South Institute, Ottawa.

Eurodad (2013). *A Dangerous Blend? The EU's Agenda to "Blend" Public Development Finance with Private Finance*. eurodad.org/files/pdf/527b70ce2ab2d.pdf/. Retrieved March 2018.

Evans, O. (2018). Connecting the poor: The Internet, mobile phones and financial inclusion in Africa. *Digital Policy, Regulation and Governance* 20:6, 568–581.

Eyben, R. and Savage, L. (2013). Emerging and submerging powers: Imagined geographies in the new development partnership at the Busan fourth High Level Forum. *The Journal of Development Studies* 49:4, 457–469.

Friederici, N. (2018). Innovation hubs in Africa: Assemblers of technology entrepreneurs. In Tynnhammer, M. (ed) *New Waves in Innovation Management Research*, 435–454. Verson Press, Wilmington.

Friederici, N., Ojanperä, S. and Graham, M. (2017). The impact of connectivity in Africa: Grand visions and the mirage of inclusive digital development. *The Electronic Journal of Information-Systems in Developing Countries* 79:1, 1–20.

Halme, M. and Korpela, M. (2014). Responsible innovation toward sustainable development in small and medium-sized enterprises: A resource perspective. *Business Strategy and the Environment* 23:8, 547–566.

Hooli, L. (2016). *Adaptability, Transformation and Complex Changes in Namibia and Tanzania: Resilience and Innovation System Development in Local Communities*. PhD dissertation, Department of Geography and Geology, University of Turku.

Hooli, L. and Haaranen, A. (2019). What is the development in the "private sector in development" approach – Perspectives from Finnish enterprises. Submitted manuscript.

Hooli, L. and Jauhiainen, J. (2017). Development aid 2.0 – towards innovation-centric development so-operation: The case of Finland in southern Africa. In Cunningham, P. and Cunningham, M. (eds) *IST-Africa 2017 Conference Proceedings*, 1–9. IIMC International Information Management Corporation, Windhoek, Namibia.

Hooli, L., Jauhiainen, J. and Lähde, K. (2016). Living labs and knowledge creation in developing countries: Living labs as a tool for socio-economic resilience in Tanzania. *African Journal of Science, Technology, Innovation and Development* 8:1, 61–70.

Hossain, M., Simula, H. and Halme, M. (2016). Can frugal go global? Diffusion patterns of frugal innovations. *Technology in Society* 46, 132–139.

ITU (International Telecommunications Union) (2019). *Data on the Internet Use and Mobile Phone Subscriptions in Africa*. www.itu.int/. Retrieved June 2019.

Jeppesen, S. (2005). Enhancing competitiveness and securing equitable development: Can small, micro, and medium-sized enterprises (SMEs) do the trick? *Development in Practice* 15:3–4, 463–474.

Jumia (2019). *Jumia*. www.jumia.com.ng/. Retrieved June 2019.

Kindornay, S. and Reilly-King, F. (2013). Promotion and partnership: Bilateral donor approaches to the private sector. *Canadian Journal of Development Studies* 34:4, 533–552.

Kolk, A. and Van Tulder, R. (2006). Poverty alleviation as business strategy? Evaluating commitments of frontrunner multinational corporations. *World Development* 34:5, 789–801.

Kraemer-Mbula, E. and Wamae, W. (2010). Adapting the innovation systems framework to Sub-Saharan Africa. In Kraemer-Mbula, E. and Wamae, W. (eds) *Innovation and the Development Agenda*, 65–90. OECD, Paris.

Leliveld, A. and Knorringa, P. (2018). Frugal innovation and development research. *European Journal of Development Research* 30:1, 1–16.

Mawdsley, E. (2015). DFID, the private sector and the re-centring of an economic growth agenda in international development. *Global Society* 29:3, 339–358.

Mawdsley, E. (2017). Development geography 1: Cooperation, competition and convergence between 'North' and 'South'. *Progress in Human Geography* 41:1, 108–117.

McEwan, C., Mawdsley, E., Banks, G. and Scheyvens, R. (2017). Enrolling the private sector in community development: Magic bullet or sleight of hand? *Development and Change* 48:1, 28–53.

Mulas, V., Nedayvoda, A. and Zaatari, G. (2017). *Creative Community Spaces: Spaces that are Transforming Cities into Innovation Hubs*. World Bank, Washington, DC.

Ndung'u, N. (2018). The M-Pesa technological revolution for financial services in Kenya: A platform for financial inclusion. In Lee, D. and Deng, R. (eds) *Handbook of Blockchain, Digital Finance, and Inclusion. Volume 1. Cryptocurrency, FinTech, InsurTech, and Regulation*, 37–56. Academic Press, Cambridge, MA.

Rasche, A., Morsing, M. and Moon, J. (eds) (2017). *Corporate Social Responsibility: Strategy, Communication, Governance*. Cambridge University Press, Cambridge.

Rowden, R. (2011). *India's Role in the New Global Farmland Grab*. GRAIN and Economics Research Foundation, New Delhi.

Scheyvens, R., Banks, G. and Hughes, E. (2016). The private sector and the SDGs: The need to move beyond 'business as usual'. *Sustainable Development* 24:6, 371–382.

Schulpen, L. and Gibbon, P. (2002). Private sector development: Policies, practices and problems. *World Development* 30:1, 1–15.

Schwartz, A., Chang, J. and Lee, M. (2014). Instrumentation and innovation in design experiments: Taking the turn towards efficiency. In Kelly, A., Lesh, R. and Baek, J. (eds) *Handbook of Design Research Methods in Education: Innovations in Science, Technology, Engineering, and Mathematics Learning and Teaching*, 47–67. Routledge, London.

Singapore (Singapore Venture Capital & Private Equity Association) (2018). www.svga.org.sg/. Retrieved April 2019.

Suri, T. and Jack, W. (2016). The long-run property and gender impacts of mobile money. *Science* 354:6317, 1288–1292.

Taylor, I. (2016). Dependency redux: Why Africa is not rising. *Review of African Political Economy* 43:147, 8–25.

Tomlinson, B. (ed) (2012a). *Aid and the Private Sector: Catalysing Poverty Reduction and Development?* IBON International, Quezon City.

Tomlinson, J. (2012b). Cultural imperialism. In Ritzer, G. (ed) *The Wiley-Blackwell Encyclopedia of Globalization*, 366–375. Wiley-Blackwell, Oxford.

UNDP (United Nation Development Programme) (2012). *Strategy for Working with the Private Sector*. www.undp.org/content/dam/undp/library/corporate/Partnerships/Private%20Sector/UNDP-Private-Sector-Strategy-final-draft-2012.pdf/. Retrieved March 2019.

Van Hove, L. and Dubus, A. (2019). M-PESA and financial inclusion in Kenya: Of paying comes saving? *Sustainability* 11, 568.

Wee Tracker (2018). *Decoding Venture Investments in Africa*. Wee Tracker.

6

UNIVERSITIES

Introduction

Universities are important for innovation activities and innovation systems. In the era of the global knowledge economy, information processing and innovation generation and exploitation, it is difficult to imagine innovation processes without universities, or at least people who have gained their education and some of their skills at universities.

Universities often deal with complex science and basic research based incremental innovations and technology development. However, in the early 21st century, it is increasingly expected that universities are societally important and that they also develop innovations with commercial value. In particular, such activities are applied tasks of entrepreneurial universities. They act as catalysts for economic development and growth within a geographic, innovation or entrepreneurship ecosystem (Gianiodis et al., 2016; De Jager et al., 2017; Etzkowitz, 2017). Entrepreneurial universities appeared in the Global North in the 1990s and began emerging in Africa in the early 21st century. Indigenous knowledge is unique to various areas of Africa, and much of its potential is still to be explored for economic growth and more socially participatory and just societies. In addition, the potential of universities as developmental agents for inclusive innovation systems is still to be fully discovered (Arocena et al., 2017).

The universities in Africa and those in the Global North have general similarities but also major differences. The basic similarities are in their role in enhancing the population's knowledge, especially at the undergraduate level. Universities provide trained and skilled people for national and regional public administrations and businesses. The main differences between the universities of the Global North and those of the Global South are in their research performance and PhD training, particularly regarding the highest performing universities' outputs.

In Africa, compared with the universities in the Global North, universities have weaker global relevance in scientific research and projects. However, Cloete and Maassen (2015) stated that whereas in the Global North countries many public and private sites and organisations produce scientific and applied knowledge, in Africa, universities are the only such sites. However, in Africa, universities are still less important in societally engaged research, business development such as university start-ups and entrepreneurial activities. Universities frequently have very few resources to develop themselves systematically. In general, the gap between universities in the more developed and less developed countries, but also within African countries, is wide. However, international development aid cooperation and projects support African universities and their capacity-building and innovation generation. African countries need to absorb available relevant knowledge, use it for their smart specialisation and build comparative advantages in those areas that have higher sustainable economic growth prospects. The use of novel technologies is vital to address the most pressing environmental and social challenges (see World Bank, 2008). Universities should take on a more impactful role in the development of African countries.

In this chapter, we discuss the role of universities in innovation processes in Africa and how international development aid for Africa relates to universities there. After this introduction, in Section 6.2, we present the general development of universities in Africa and their situation regarding innovation-related activities and innovation systems. Later, in Section 6.3, we discuss the emergence and different trajectories of entrepreneurial universities, technological hubs and innovation districts in Africa. In Section 6.4, we explore international development aid that focuses on universities in Africa, especially support for their innovation processes and activities. We also present concrete cases of such support deriving from the Global North to African universities, as well as South-South collaboration. In Section 6.5, we present the key conclusions of the chapter.

- Universities have a key role in innovation systems and innovation processes, but African universities have been weaker in innovation-related activities.

- Very few universities in Africa outperform in globally significant research or are entrepreneurial in their innovation-related activities.
- Some international development aid cooperation and projects aim to enhance capacity building and innovation generation in African universities.

Development of innovation-oriented universities in Africa

African universities have different development trajectories. In general, most key universities established before the 1960s' wave of independence in Africa were affiliated with partner universities in the colonising countries, such as the United Kingdom, France and Portugal. They were less-qualified extensions of their university systems with the task to train the tiny elite of the colonies. Their educational qualification system was close to that of their European counterparts at that time (Materu et al., 2011; Cloete & Maassen, 2015). Typical of this era was the universities' focus on academic studies and excellence, without major engagement in the economic development of the surrounding communities and societies or innovation-focussed activities.

After their independence, many African countries formulated their first strategies to enhance science and technology. They established universities and research institutes following the examples of the OECD countries where these institutions were well developed (Clark & Frost, 2016). In the newly independent African states, authorities became more involved in university development. Universities were important sites to improve human resources in fields considered significant for each independent African country. With a pan-African enthusiasm, universities were to be developmental agents to be steered by national governments, as declared in 1962 by the members of the Association of African Universities and in the conference of higher education development in Africa, organised by UNESCO and UNECA and held in 1962 in Tananarive, Madagascar (Coleman, 1986; Cloete et al., 2015). Universities were not only institutions of academic novelty, but also sites for potential political activism through continuous inflow of young generations motivated to change society. Therefore, in some cases, the state authorities regulated the main activities of universities to fit them better to the political development of respective African countries. Furthermore, by the early 1970s, the number of research institutes in Africa rose quickly from a couple of hundred to over 2,000. They employed about 11,000 persons, with an average of 5.5 employees per institute (AMCOST, 2006). Many institutes were thus very small and could not play important developmental roles in their respective countries.

The rapid population growth and the need to elevate the general education levels of the African population meant that the amount of students grew continuously. Materu et al. (2011) stated that in Sub-Saharan Africa in 1985–2002, the number of tertiary students increased from 800,000 to about 3 million, growing by about 15% per year on average. At the same time, the public expenditure to the universities grew at a substantially slower pace. This created a challenge for the provision of good-quality and individualised education. In addition, the resources available had to be directed increasingly towards fulfilling educational goals instead of investing in high-level research (see Wiseman & Wolhuter, 2013). Behind such twists and turns in higher education policy were also suggestions from global organisations, such as the World Bank, to focus more on primary than on tertiary education with university development (Cloete & Maassen, 2015). As a result, during the 1980s and 1990s, investments in science and technology were not prioritised in Africa, and in reality, they were isolated, underfunded and not linked to other development institutions.

During the 1990s, the initiation of restructuring of the university system led to the privatisation of many universities, which further diversified the universities' roles and possibilities in society and innovation for development. Many new private universities responded to the growing need for higher education. Some of them remained under-resourced and with a rather low teaching quality and poor performance of their students. Other private universities developed with donations from foreign countries and advanced (see Salmi, 2009). Later in the 21st century, some agile private universities attracted highly skilled students and researchers and became important sources of new basic and applied knowledge. Privatisation can lead to an advantage in developing more agile practices for innovation in universities, but it can also be a tool to attract less qualified student cohorts and pass them through the higher education system.

The early 21st century witnessed a shift in Africa in the development policies regarding universities. Closely following the development discourse in the Global North, and the key international actors such as the World Bank, African countries saw universities as engines for economic growth through commercial innovations. Investments in tertiary education, including universities, raised the national GDP, which legitimised the integration of universities into neoliberal economic policies (McCowan, 2016). This also paved the way to engage universities and their innovation-oriented scholars as partners in international development cooperation between the Global North donor countries and universities in selected African countries. Such a policy turn facilitated the emergence of entrepreneurial universities in Africa. Overall, according to McCowan (2016), it is about commodification of knowledge and universities in which teaching, research and knowledge are transformed into simple packages for sale and profit-making.

Although many public African universities gained more autonomy, their financial autonomy remained limited (i.e. they substantially depended on public expenditure). Other important sources were students' fees, endowments, grants and internally generated revenues (Materu et al., 2011). The role of external funding started to increase, including contracted research, international cooperation and donations. This resonates with the need for African universities to be connected with the rapidly developing economy and society, and in particular with innovation networks and ecosystems (see Cloete & Maassen, 2015). However, in this changing higher education landscape, persisting tension surrounded the purpose of a public university: Should it produce global-level research, prepare a skilled labour force or address societal inequalities (Swartz et al., 2019)?

Currently on the agenda of many African universities are an increased focus on research-driven activities, postgraduate education and internationalisation of universities' activities, including research and education (for a case study of one African university, the University of Ibadan in Nigeria, see Materu et al., 2011; for several African universities, see Teferra 2017a). In addition, many private universities have emerged in both education and research fields. The recent general development of African universities is characterised by the growth in the number of students and universities as well as by increasing connection to global knowledge flows, national economic development and innovation systems.

Much of the recent emphasis on African universities as a major potential generator of innovations has tended to focus merely on economic issues. The business-oriented role of higher education in Africa has taken priority over universities' potential for emancipation of communities and individuals (Scott, 2011; Diamini, 2018). Such a narrow perspective undermines the fact that universities play an important potential role in the democratisation of African societies, providing better social equality and opportunities through open innovation systems and as platforms of continuous transformation and search for novelty. As McCowan (2016: 136) stated, "the worth of the goods provided by the university would be in the tangible benefits brought to society". Already in the early 1990s, Kerr (1991) mentioned that aside from the economy-oriented production and consumption deriving from the main research, education and service tasks of universities, universities are also important in broader citizenship issues through socialisation of students, fostering the democratisation principles and practices of society and critical evaluating developments when needed. Taking indigenous knowledge as a support for development is not yet very common in African universities. However, deeper engagement with such issues can provide a competitive advantage for innovation for development in Africa compared with many universities in economically more advanced countries, as well as in their engagement towards more participatory and socially just societies and communities.

The discussion of the development of universities and their role in African society is linked to the number and performance of universities in Africa. It is difficult to estimate the exact number of universities in Africa. In 2019, the Cybermetrics Lab (2019) put in ranking 1,685 African universities. Some universities do not appear on this list, and every year, new private and public universities are established. Behind this is the demographic growth of African countries and the need to expand higher education in their trajectories towards more industrial and economically advanced societies.

In several African countries, there are tens of universities; however, their scope and quality vary. According to the most-used global ranking of universities (see ARWU, 2018), of the top 100 highest ranking universities globally, 95 originate from somewhere other than the Global South. No African universities feature among the top 100 universities. Of the top 500 universities, 88 are from outside the Global North and five of them are from Africa. Therefore, only about 1% of the top 500 universities in the world originate from Africa. This share is small compared with the share of the African population, which is 16.7% of the global population. Furthermore, because it is estimated that the global share of the African population will grow to 20% by 2030, there is a strong need to develop universities in Africa. Africa still lags substantially behind in R&D investments, number of patents, scientific publications and researchers – the main factors supporting knowledge-based development (UNECA, 2018).

The geography of the highest-performing universities in Africa is very uneven. Based on several rankings, the top 25 universities in Africa are presented in Table 6.1. Africa has 54 independent countries, but of the continent's top 12 universities, nine are from South Africa, two from Egypt and one from Kenya, and among the top 25 are universities from seven countries (Table 6.1; see also Figure 1.4). Lower in the rankings, more countries appear on the list. In the top 100 are universities from 19 African countries. Those universities positioned lowest on this top 100 list are ranked among the 3,000–4,000 best universities globally (Cybermetrics Lab, 2019). This also indicates that the global academic competitiveness of most African universities, measured with conventional performance indicators, is weak. However, universities in Africa as well as in the Global South have so far only seldom been clearly research-oriented universities (Altbach, 2013). This partially explains their lower rankings because research performance, with its various outputs, is one important indicator in evaluating universities. Therefore, a more feasible target for enhancing higher education could be flagship universities that, according to Douglass (2016), provide education and are research-intensive, but have wider local societal roles than the top-level universities. Several flagship universities have been identified as capacity-builders and trendsetters in the academic life of different African countries; however, they also face many challenges in the changing higher education markets in the continent (Teferra, 2017b).

TABLE 6.1 Top 25 universities in Africa

Name of university	Country	Location
1 University of Cape Town	South Africa	Cape Town
2 University of Witwatersrand	South Africa	Johannesburg
3 Stellenbosch University	South Africa	Stellenbosch
4 University of Pretoria	South Africa	Pretoria
5 Cairo University	Egypt	Cairo
6 University of KwaZulu-Natal	South Africa	Durban
7 University of Johannesburg	South Africa	Johannesburg
8 Alexandria University	Egypt	Alexandria
9 University of the Western Cape	South Africa	Cape Town
10 University of Nairobi	Kenya	Nairobi
11 North West University	South Africa	Potchefstroom
12 University of South Africa	South Africa	Pretoria
13 Makerere University	Uganda	Kampala
14 Mansoura University	Egypt	Dahaklia
15 Ibadan University	Nigeria	Ibadan
16 American University in Cairo	Egypt	Cairo
17 Ain Shams University	Egypt	Cairo
18 Rhodes University	South Africa	Grahamstown
19 University of Ghana	Ghana	Accra
20 Benha University	Egypt	Banha
21 Beni-Suef University	Egypt	Beni Suef
22 Mohammad V University of Rabat	Morocco	Rabat
23 Covenant University	Nigeria	Ota
24 Tshwane University of Technology	South Africa	Pretoria
25 University of Cadi Ayaad	Morocco	Marrakesh

Source: ARWU (2019); Cybermetrics Lab (2019), The QS (2019)

Entrepreneurial universities, innovation districts and hubs in Africa

In general, the commercialisation of knowledge and research generated by universities are globally common activities (Grimaldi et al., 2011). An entrepreneurial society refers to the context in which knowledge-based entrepreneurship is a driving force for economic growth, employment creation and competitiveness. In this context, an entrepreneurial university is a significant institution for knowledge production and dissemination (Guerrero & Urbano, 2012). The aim of entrepreneurial universities is to stimulate economic development, enhance competitiveness, increase wealth creation and produce innovations (i.e. to fulfil the third mission of universities). Currently, many African universities show the interest and motivation to become entrepreneurial. At the end of the 2010s, it was still rare to find an African university with a strong mission, vision, strategy and practice for entrepreneurialism. The very scarce research on the topic focusses on African universities' student entrepreneurship (Amadi-Echendu et al., 2016; De Jager et al., 2017) and not on the entrepreneurial university itself (see Doh et al. 2019).

African universities have different motivations, goals and trajectories in becoming entrepreneurial. Some wish to enhance their students' entrepreneurial mindsets, creativity and employability. With the expansion of the population in tertiary and university education, it is important for graduating students to find employment based on their skills. Other universities aim to generate additional income because many universities in Africa suffer from global financial austerity and limited national resources (Johnstone, 2003; Oketch, 2016). Many entrepreneurially oriented universities aim to improve their response to national and regional economic development, business relationships between the university and the surrounding private sector and engagement with national innovation systems and business ecosystems. Entrepreneurial orientation of universities should lead to reduction of poverty in their respective countries.

Various African countries undertake university reforms towards increased commercialisation of research results. Universities set up technology transfer offices, incubators, entrepreneurship centres and internal seed funds (Rasmussen & Borch, 2010). Nevertheless, there are different pathways for their strategies, policy-making and broader holistic transformation to become entrepreneurial, as well as differences in their performance in their entrepreneurship-related activities. Universities alone do not decide if, when and how they become entrepreneurial. Grimaldi et al. (2011) explained that entrepreneurial strategies of universities are influenced by many factors, including government policies enacted by local, regional, national and supranational actors; the culture of individual universities and their subunits; individual campus leadership; the quality of the university; the resources and dynamism of the local economy; and capabilities to transfer knowledge and technology. Many aspects need to be considered when developing entrepreneurial universities in African contexts.

Despite the fact that there are only a few entrepreneurial universities in Africa, a major difference exists in the structure of entrepreneurial universities between Anglophone and Francophone Africa. According to Doh et al. (2019), in Anglophone African countries, the entrepreneurial aspect of the higher education and university system consists of applied, technological and vocational institutions. It is a very interactive and comprehensive subsector within those universities aiming to be entrepreneurial. For this, many universities need to undergo significant governance transformation. In Francophone African countries, the entrepreneurial missions of universities are mostly driven by specific applied establishments, the function of which is to be entrepreneurial. Other Francophone universities in Africa have, in general, less interest in entrepreneurial activities and less evidence of being entrepreneurial.

BOX 6.1 ENTREPRENEURIAL UNIVERSITIES

The concept of entrepreneurial universities varies (Clark, 1998; for Africa, see Doh et al. 2019). One major scholar of the topic, Burton Gibb (2012), distinguishes between the concepts of entrepreneurship, enterprising activities of universities and entrepreneurial universities. An entrepreneurial university is any university that contributes and provides leadership for creating entrepreneurial thinking, actions, institutions and capital. Such activities reach beyond common technology transfer with patents, spin-offs and start-ups created by other universities (Audretsch & Keilbach, 2007). With broader leadership, an entrepreneurial university enhances competitive economic growth, creates wealth both for the university and society and increases the university's impact within and beyond the region in which it is located (Guerrero et al., 2015). Academic staff, administrators, students, existing firms and new ventures are key elements of entrepreneurial universities (Etzkowitz & Zhou, 2008; Gjerding et al., 2008). At entrepreneurial universities, innovations developed by these actors are key instruments for achieving broader impacts. However, McCowan

> (2016) saw a risk in commodification of knowledge and universities when it becomes entrepreneurial and oriented towards business.
>
> In their study, Pugh et al. (2018) found that entrepreneurial universities and their entrepreneurship departments have formal and informal roles in the development of the surrounding society. The formal roles are performed through collaborative research, contract research and consulting, both via direct links to the nearby region and through contribution to the wider structure. The informal roles are ad hoc advice provision and practitioner networking, mainly via individuals' direct connections to the surrounding region (see also Perkmann et al., 2013).

Entrepreneurial universities are usually connected to technological hubs and innovation districts. These are integral to the innovation-related scene in Africa in the early 21st century. There are technological hubs and/or innovation districts in most African countries, especially in their capitals and other larger urban agglomerations. However, many bottom-up innovations emerge outside of them from poorer areas, as discussed in Chapter 8. Nevertheless, technological hubs and innovation districts are tools specifically designed to develop innovations.

Conceptually, similarities and differences exist between technological hubs and innovation districts. Both use many kinds of knowledge: synthetic, applied and symbolic (see Chapter 2). Hubs tend to be smaller sites focussed on technology start-up development, experimentation and networking. Usually, hubs refer to concrete buildings and sites, but in their broader definition, they cover large high-technology regions such as Silicon Valley in the United States or Innovation City in Kigali, Rwanda, Yabacon Valley in Nigeria and Cape Town in South Africa. Broadly speaking, these large regions are the clustering settings for smaller hubs.

Innovation districts are broader forms of urban development where firms, organisations and institutions, including universities, connect across disciplines to create a centre for innovation. An environment with access to amenities, ability to mix with other related innovation actors and smaller walkable size to enable meeting each other is important (Thuesen Pedersen, 2018, see Table 6.2). Specifically designed innovation districts usually have a broader diversity of research and institutions supporting research, enterprises and start-ups in various fields of technology and business development, creating an interactive innovation ecosystem. Innovation districts exist in many African countries, such as South Africa (Box 6.2).

TABLE 6.2 Key aspects in the promotion of innovation districts

1. Clustering of innovative sectors and research strengths is the backbone of innovation districts.
2. For innovation districts, convergence – the melding of disparate sectors and disciplines – is king.
3. Districts are supercharged by a diversity of institutions, companies and start-ups.
4. Connectivity and proximity are the underpinnings of strong district ecosystems.
5. Innovation districts need a range of strategies – large and small moves, long-term and immediate.
6. Programming is paramount. Programming – a range of activities to grow skills, strengthen firms and build networks – is the connective tissue of a district.
7. Social interactions between workers – essential to collaboration, learning and inspiration – occur in concentrated "hot spots".
8. Make innovation visible and public.
9. Embed the values of diversity and inclusion in all visions, goals and strategies.
10. Get ahead of affordability issues.
11. Innovative finance is fundamental to catalysing growth.
12. Long-term success demands a collaborative approach to governance.

Source: Wagner et al. (2017)

BOX 6.2 STELLENBOSCH INNOVATION DISTRICT, SOUTH AFRICA

Stellenbosch Innovation District (SID) features one leading African university, the University of Stellenbosch in South Africa. SID's aim is to support the development of innovative solutions for urban, social, economic and environmental challenges and to transform the rapidly growing town of Stellenbosch (180,000 inhabitants) into a smart and sustainable urban settlement with a collaborative culture, inclusion and connections between different communities.

Behind the idea of SID was the development of the assistance programme COFISA (Co-operation Framework on Innovation Systems between Finland and South Africa) from Finland to Africa. The programme found that to enhance national innovation development in South Africa, regional innovation system strategies should be implemented. Stellenbosch was selected as one site, and in 2012 launched the development of a local innovation ecosystem. International, national and local experts and stakeholders held meetings, and in 2014, its strategy (turning challenges into opportunities through connecting, innovating and sustaining) was published. However, local politics and bureaucracy hindered SID's development (van Heyningen, 2014).

SID has supported projects such as iShack (solar panel service for underserved communities), Sustainable Brothers and Sisters (platform for sustainable living strategies through research, collaboration and networking) and Qurio.co (opinions from audience, feedback from employees, measuring customer satisfaction and evaluating workshop audience). Several other projects were also launched, but many remained active only for a short time.

International development cooperation for universities

As discussed previously, the World Bank and other international organisations suggested in the 1990s that the enhancement of primary education is the key for development in Africa. Such a perspective resulted in only very limited international aid for university development in Africa. However, beginning towards the end of the 1990s, and culminating in the early 21st century, the World Bank's viewpoint changed completely. Universities were seen as key factors for positive development in Africa. Universities produce skilled workers who are able to absorb and develop relevant technologies for the diversification of the economy, especially the growing industry that is still in a rather initial stage in many African countries (MacGregor, 2015).

When the new understanding made a breakthrough, international development aid was soon targeted towards selected African universities. For example, the Swedish development agency SIDA started a long-term development cooperation with Makerere University in Uganda. Such bilateral cooperation helped to substantially increase Makerere's research and PhD training performance (Box 6.3). In Finland, the Ministry for Foreign Affairs started to provide funding for the Higher Education Institutions Institutional Cooperation Instrument (HEI ICI) through its development cooperation funds. The idea was to enhance higher education provision and capacity-building in selected Global South countries, including those in Africa. The programme was organised through cooperation projects between higher education institutions of the donor country Finland and higher education institutions in the aid-receiving countries. The aim was to develop their subject-specific, methodological, educational and administrative capacity. It was seen that support for education helps to promote skills development; establish a well-functioning, efficient and equal society; encourage entrepreneurship; drive sustainable development and reduce poverty (Finnish National Agency for Education, 2019).

BOX 6.3 SWEDISH SUPPORT FOR CAPACITY BUILDING AT MAKERERE UNIVERSITY, UGANDA

The Swedish International Development Cooperation Agency (SIDA) and Makerere University in Uganda started a collaborative partnership in 2000. SIDA supported the democratic governance at public universities and the analytical capacity improvement of the research on poverty reduction (SIDA, 2018).

The main funding was for Makerere University in Kampala and included a start-up business incubator, libraries, quality assurance and laboratory equipment and administrative reforms for effective research management. In addition, four other public universities in Uganda were supported. The focus was on supporting postgraduate education within medicine, technology, humanities and social sciences, agriculture and veterinary medicine. The scholars spent periods in Uganda and at Swedish universities. Particularly, capacity-building was a method to create competence in the supervision of PhD and master's students. SIDA's support ended in 2020, which created challenges in Makerere to continue with the same number of activities. Over 20 years of the programme, more than 350 academic staff completed PhDs, 250 earned master's degrees and 75 postdoctoral researchers completed fellowships. Out of a total of 668 scholarships, 278 (42%) went to women, and 8,000 papers were published in local, regional and international journals (Nakkazi, 2019). Makerere university is currently firmly based among the top 20 universities in Africa.

SIDA also supported many other activities in Uganda. Among these were business initiatives through the global programme Innovations Against Poverty (IAP), improvement of the agricultural business and investment climate, increase of trade and cooperation with the private sector. A particular target has been to lower the prevalence of HIV in Uganda (SIDA, 2018).

During the 2010s, criticism towards the aid emerged in Sweden due to widespread corruption and legal oppression against sexual minorities in Uganda. Due to this, Sweden suspended its aid to Uganda for one year. In 2018, SIDA's support to Uganda was $56.6 million USD (SIDA, 2018).

When the understanding of the prominent role of universities started to prevail in the countries of the Global North, the capacity building of African universities became an important topic in the international development cooperation and aid for Africa. There is a growing demand for quality higher education in Africa, so many non-African countries are actively present in the African education market. They aim to meet the demand for education while also making a profitable business. Education export became a huge and continuously growing multibillion business in the 21st century, especially for the United Kingdom, the United States and Australia. There were different strategies and motivations for entering into the African education market (Box 6.4).

BOX 6.4 EXPORT OF EDUCATION EXPERTISE FROM FINLAND TO AFRICA

Finland is a newcomer to education export activities. Due to its rather late appearance in this market, the Finnish government and the related actors have taken a slightly different perspective than the mainstream education export providers in Africa. Finland stresses the notion of export of education expertise because aside from educational services, the Finnish offering includes consulting services and technological solutions for facilitating learning processes. Furthermore, supported by the high position of Finland in various education-related rankings, the Finnish education providers are expected to possess modern and future-oriented concepts of learning, learning environments and pedagogy (Juntunen, 2014). To date, Finland has exported education to 10 countries in Africa with a value of 350 MEUR. The amount is globally very small; however, this export is also connected to a broader innovation-focussed development aid policy of Finland (see Chapter 4).

As a novel example, in 2019, the University of Turku (UTU, Finland) opened a campus at the University of Namibia (UNAM) in Windhoek, Namibia. There, a student can receive a master's degree that is accredited by UTU. Initially, the themes related to software technology and engineering. The scope of the campus extends to broader innovation-related activities, and start-up companies and spin-offs emerge. In cooperation with UNAM, the government of Namibia and the private sector, the campus is developing into a technology hub that is important for Namibia's national innovation system.

The prevailing form has been to export education and knowledge from the Global North to the Global South. However, the amount of South-South education exports is rising as well. In addition, several African countries are looking for possibilities to export education to other African countries. However, not all education, training and capacity building required large programmes and institutional settings. In the field of innovation and development studies, the African Network for Economics of Learning, Innovation, and Competence Building Systems (AfricaLics) was established. Its mission is to function as a network between scholars, policymakers and other stakeholders to promote inclusive and sustainable social and economic development through better understanding of the role of innovation in development (Box 6.5).

BOX 6.5 AfricaLics

AfricaLics connects scholars working with a specific focus on innovation and development in African countries. It is a regional dimension of Globelics, a worldwide, open and diverse network community of scholars working on innovation and competence-building in the context of economic development. In Globelics, engaged scholars have different backgrounds in scientific disciplines, age, gender and national origin. The participants promote inclusive and sustainable social and economic development and are interested in how innovation and competence building link to economic development. There is a strong idea to connect the innovation-focussed scholars from the Global North, the Global South and beyond.

The background of AfricaLics stemmed from the need to understand and provide capacity-building from theoretical and practical perspectives on innovation and learning in the context of Africa. Similarly to the Globelics organisation, AfricaLics increases networking opportunities and access to education to enhance economic and socially sustainable development in Africa. It facilitates the production and use of high-quality research in innovation and development with a view to promoting inclusive and sustainable development in African countries.

To achieve these objectives, the network organises various activities such as online networking platforms to encourage discussion and linkages between researchers working in African innovation and development; annual conferences to gather scholars located and working on innovation and development in Africa; PhD academies, such as a two-week research training for second- and third-year PhD students studying at African universities; visiting fellowships as opportunities for PhD and postdoctoral students from low- and low-middle-income African countries to receive additional training and capacity-building support; and promotion of research activities, research projects and dedicated innovation and development training at African universities (Africalics, 2019).

Conclusions

In Africa, as everywhere, universities are key actors in knowledge generation and innovation development. A particular challenge in Africa is the generally rather low quality of universities if they aim to become world-class research-based universities. Currently, only a handful of African universities can clearly claim that status based on their performance. The main role for the majority of universities in Africa has been to provide higher education opportunities for a large population. However, in the 21st century, policies and practices in which universities aim to become more involved in innovations, businesses and other societally relevant activities have emerged. Universities are finding their place in national and regional innovation systems. Also, entrepreneurial universities are emerging aiming to create commercially successful innovations.

International development aid projects have supported African universities and capacity-building, sometimes even for decades. The most recent aid is development cooperation, in which the focus is often on technology development and creation of various kinds of innovation hubs. Several African countries continue to progress quickly, so Global North countries are also withdrawing their long-term support from universities.

In the 2020s, African universities will face challenges and opportunities to be involved in innovation for development. The rapid population growth and the delivery of higher education to the growing youth creates pressures to manage such expansion. The global competition in research is strong, so to break through to the top 100 or even top 500 universities globally requires huge investments in human skills, infrastructure and research practices in African universities.

A potential opportunity not yet much discovered in African universities is indigenous knowledge. It can link local communities (see Chapter 8), universities and economic agents to innovations with a competitive advantage.

Along with the attempts for global competitiveness and relevance, African universities need to be actively involved in local development processes and be progressive in their attempts to help solve poverty and inequalities that are still major development challenges across the continent.

- Some African universities foster their entrepreneurial characters to create better employment opportunities for their students, generate additional income and have more societal relevance.
- African universities need to be actively engaged with local development processes with locally relevant innovations.
- The international development cooperation for African universities is changing from broader capacity building into narrower and more specific tasks and technology-related activities.

Discussion questions

- Where are the best-performing universities in Africa and why are they in these countries?
- What opportunities and challenges exist for the development of entrepreneurial universities in Africa?
- Discuss how universities in Africa could be better incorporated into national innovation systems and local development.

References

Africalics (2019). *Africalics*. www.africalics.org/. Retrieved June 2019.

Altbach, P. (2013). Advancing the national and global knowledge economy: The role of research universities in developing countries. *Studies in Higher Education* 38:3, 316–330.

Amadi-Echendu, P., Chodokufa, K. and Visser, T. (2016). Entrepreneurial education in a tertiary context: A perspective of the University of South Africa. *International Review of Research in Open and Distance Learning* 17:4, 22–34.

AMCOST (African Ministerial Council on Science and Technology) (2006). *AMCOST Extraordinary Conference*. Cairo and S&T Consolidated Plan of Action, November 20–24.

Arocena, R., Göransson, B. and Sutz, J. (2017). *Developmental Universities in Inclusive Innovation Systems. Alternatives for Knowledge Democratization in the Global South*. Springer, Berlin.

ARWU (2018). *Academic Ranking of World Universities*. www.shanghairanking.com/ARWU2018.html/. Retrieved June 2019.

Audretsch, D. and Keilbach, M. (2007). The theory of knowledge spillover entrepreneurship. *Journal of Management Studies* 44:7, 1242–1254.

Clark, B. (1998). *Creating Entrepreneurial Universities: Organizational Pathways of Transformation*. Pergamon Press, New York.

Clark, N. and Frost, A. (2016). It's not STI: It's ITS – the role of science, technology and innovation (STI) in Africa's development strategy. *International Journal of Technology Management & Sustainable Development* 15:1, 3–13.

Cloete, N., Bunting, I. and Maassen, P. (2015). Research universities in Africa: An empirical overview of eight flagship universities. In Cloete, N. and Maassen, P. (eds) *Knowledge Production and Contradictory Functions in African Higher Education*, 18–31. African Minds, Cape Town.

Cloete, N. and Maassen, P. (2015). Roles of universities and the African context. In Cloete, N. and Maassen, P. (eds) *Knowledge Production and Contradictory Functions in African Higher Education*, 1–17. African Minds, Cape Town.

Coleman, J. (1986). The idea of developmental university. *Minerva* 24:4, 476–494.

Cybermetrics Lab (2019). *Ranking Web of Universities*. www.webometrics.info/. Retrieved June 2019.

De Jager, H., Mthembu, T., Ngowi, A. and Chipunza, C. (2017). Towards an innovation and entrepreneurship ecosystem: A case study of the Central University of Technology, Free State. *Science, Technology and Society* 22:2, 310–331.

Diamini, R. (2018). Corporatisation of universities deepens inequalities by ignoring social injustices and restricting access to higher education. *South African Journal of Higher Education* 32:5, 54–65.

Doh, P., Jauhiainen, J. and Boohene, R. (2019). Entrepreneurship patterns and the entrepreneurial university transformation in Africa: Gaps and synergistic potentials. Submitted manuscript.

Douglass, J. (2016). Profiling the new flagship university model. In Douglass, J. (ed) *The New Flagship University: Changing the Paradigm from Global Ranking to National Relevancy*, 39–102. Palgrave Macmillan, Basingstoke.

Etzkowitz, H. (2017). Innovation lodestar: The entrepreneurial university in a stellar knowledge firmament. *Technological Forecasting & Social Change* 123, 122–129.

Etzkowitz, H. and Zhou, C. (2008). Introduction to special issue building the entrepreneurial university: A global perspective. *Science and Public Policy* 35:9, 627–635.

Finnish National Agency for Education (2019). *Support for the Internationalization of Higher Education*. www.beta.oph.fi/. Retrieved June 2019.

Gianiodis, P., Markman, G. and Panagopoulos, A. (2016). Entrepreneurial universities and overt opportunism. *Small Business Economies* 47, 609–631.

Gibb, A. (2012). Exploring the synergistic potential in entrepreneurial university development: Towards the building of a strategic framework. *Annals of Innovation and Entrepreneurship* 3:1, 1–24.

Gjerding, A., Wilderom, C., Cameron, S., Taylor, A. and Scheunert, K. (2008). Twenty practices of an entrepreneurial university. *Higher Education Management and Policy* 18:3, 1–28.

Grimaldi, R., Kenney, M., Siegel, D. and Wright, M. (2011). 30 years after Bayh-Dole: Reassessing academic entrepreneurship. *Research Policy* 40:8, 1045–1057.

Guerrero, M., Cunningham, J. and Urbano, D. (2015). Economic impact of entrepreneurial universities: An exploratory study of the United Kingdom. *Research Policy* 44:3, 748–764.

Guerrero, M. and Urbano, D. (2012). The development of an entrepreneurial university. *Journal of Technology Transfer* 37:1, 43–74.

Johnstone, D. (2003). Cost-sharing in higher education, tuition, financial assistance and accessibility in comparative perspective. *Czech Sociological Review* 39:3, 351–374.

Juntunen, T. (2014). Education export – what does it mean? *Journal of Universities of Applied Sciences* 3.

Kerr, C. (1991). *The Great Transformation in Higher Education*. State University, Albany.

MacGregor, K. (2015). Higher education is key to development – World Bank. *University World News*, April 10.

Materu, P., Obanya, P. and Righetti, P. (2011). The rise, fall and reemergence of the University of Ibadan, Nigeria. In Altbach, P. and Salmi, J. (eds) *The Road to Academic Excellence: The Making of World-Class Research Universities*, 195–227. World Bank, Washington, DC.

McCowan, T. (2016). Universities and the post-2015 development agenda: An analytical framework. *Higher Education* 72:4, 505–523.

Nakkazi, E. (2019). Makerere prepares for end of 20 years of Swedish support. *World University News*, January 11.

Oketch, M. (2016). Financing higher education in Sub-Saharan Africa: Some reflections and implications for sustainable development. *Higher Education* 72, 525–539.

Perkmann, M., Tartarik, V., McKelvey, M., Autio, E., Broström, A., D'Este, P. and Sobrero, M. (2013). Academic engagement and commercialisation: A review of the literature on university – industry relations. *Research Policy* 42:2, 423–442.

Pugh, R., Lamine, W., Jack, S. and Hamilton, E. (2018). The entrepreneurial university and the region: What role for entrepreneurship departments? *European Planning Studies* 26:9, 1835–1855.

Rasmussen, E. and Borch, O. (2010). University capabilities in facilitating entrepreneurship: A longitudinal study of spin-off ventures at mid-range universities. *Research Policy* 39:5, 602–612.

Salmi, J. (2009). *The Challenge of Establishing World-Class Universities*. World Bank, Washington, DC.

Scott, P. (2011). An African take on internationalization. *World University News*, November 25.

SIDA (Swedish International Development Cooperation Agency) (2018). *Our Work in Uganda*. www.sida.se/English/where-we-work/Africa/Uganda/Our-work-in-Uganda/.

Swartz, R., Ivancheva, M., Czerniewicz, L. and Morris, N. (2019). Between a rock and a hard place: Dilemmas regarding the purpose of public universities in South Africa. *Higher Education* 77:4, 567–583.

Teferra, D. (ed) (2017a). *Flagship Universities in Africa*. Springer, Berlin.

Teferra, D. (2017b). African flagship universities: Epilogue. In Teferra, D. (ed) *Flagship Universities in Africa*, 507–516. Springer, Berlin.

The QS (2019). *QS World University Rankings*. www.topuniversities.com/. Retrieved June 2019.

Thuesen Pedersen, K. (2018). Innovation districts mark the 4th industrial revolution. *COWI*. www.cowi.com/. Retrieved June 2019.

UNECA (United Nations Economic Commission for Africa) (2018). *Africa Sustainable Development Report 2018. Towards a Transformed and Resilient Continent.* UNECA, Addis Ababa.

van Heyningen, P. (2014). The opportunity for developing Stellenbosch as an innovation district. *Capeinfo* capeinfo.com/more/interviews/297-stellenbosch-delivering-as-innovation-capital-of-sa?layout=/. Retrieved June 2019.

Wagner, J., Andes, S., Davies, S., Storring, N. and Vey, J. (2017). *Metropolitan Revolution. 12 Principles Guiding Innovation Districts.* Brookings Institution. www.brookings.edu/. Retrieved June 2019.

Wiseman, A. and Wolhuter, C. (2013). *Development of Higher Education in Africa: Prospects and Challenges.* Emerald Group, Bingley.

World Bank (2008). *Accelerating Catch-up: Tertiary Education for Growth in Sub-Saharan Africa.* World Bank, Washington, DC.

7

NON-GOVERNMENTAL, INTERGOVERNMENTAL AND OTHER AID ORGANISATIONS

Introduction

Epochal shifts towards the beyond aid agenda and innovation-focussed development cooperation have brought, in the early 21st century, new actors to the international field of development and changed the role and mission of some customary organisations. Non-governmental and intergovernmental organisations have always been involved in development, as well as philanthropic non-profit foundations. However, more recently, global super-rich individuals and transnational companies have become involved in development cooperation, and they increasingly apply market-based approaches in development cooperation even if they would not be seeking profit themselves – philanthrocapitalism, as some critical scholars name it (Edwards, 2008). Sometimes such powerful interventions have been seen as new forms of global governance (Levich, 2015). New modalities and interactions in unilateral cooperation and cross-sectoral partnerships (CSPs) between different actors are an inherent part of contemporary development cooperation (Vestergaard et al., 2019).

The role of these actors in innovation-focussed development cooperation is two-fold. First, as have many other actors, they have supported the emergence of innovations and local innovation systems and development. Oftentimes, non-governmental organisations (NGOs) especially can be an assembling link between different actors such as international private companies and local communities. Second, these new non-state actors have brought innovative and novel organisational models and operative activities to development cooperation.

After this introduction, in Section 7.2, we discuss the recent developments in the NGO sector related to development aid and assistance. The activities of NGOs are connected to the broader dynamics in international development cooperation. In addition, the beyond aid agenda and its focus on for-profit innovations have especially changed the scopes and roles of NGOs in development cooperation. Innovations, entrepreneurship and cooperation with private sector actors have become more common compared to the 20th century.

In Section 7.3, we focus on intergovernmental organisations having global reach in their activities. These include the key international actors (e.g. the World Bank, IMF and UN) influencing global development, including that in Africa. After this, in Section 7.4, we discuss global not-for-profit and for-profit foundations. In Section 7.5, we illustrate examples of traditional, conventional and contemporary grassroots NGOs that might not have global goals or reach but that are very important in providing local assistance for individuals and communities in Africa. Finally, in Section 7.6, we present the conclusions of the chapter.

- NGOs and other non-public and non-private organisations have become important stakeholders and agents in international development cooperation.
- The beyond aid agenda in development cooperation also influences the strategies and practices of NGOs involved in development issues by bringing more of their attention to entrepreneurship and innovations as development goals and tools.
- Globally, large charity organisations can have even larger roles in development of the less developed countries in the Global South than the traditional development aid by the countries of the Global North.

NGOs and other developmental organisations

NGOs in development cooperation consist of a heterogeneous group of actors (see Box 7.1). In addition to public authorities, NGOs have always been involved in development cooperation. Charities and religious organisations have been active in the Global South long before the beginning of official development assistance. In the 21st century, the variety of NGOs and other non-public and non-private organisations involved in development cooperation has increased (Box 7.1). Their strategic objectives and operational practices have also been in transition.

BOX 7.1 NGOs IN INTERNATIONAL DEVELOPMENT AID

In principle, two types of NGOs exist. There are NGOs that advocate and influence governments, for example, to enhance social, cultural, economic or environmental sustainability, welfare and justice. The other type of NGOs conduct operational activities to promote the aforementioned issues and provide related services. NGOs are usually non-profit organisations. In principle, NGOs operate independently of national or intergovernmental organisations. Therefore, they do not have to necessarily follow or support the principles of other donors and aid-receiving countries. However, they are influenced at least indirectly by these organisations. For example, the UN has set certain criteria that NGOs must follow to cooperate with it, and assisted countries have various rules and regulations that NGOs need to follow.

The scopes, functions and sizes of NGOs vary. Some NGOs lack transparency and accountability in development activities, and their decisions are not necessarily made democratically. Others function more like an open platform to engage ideas and people in development cooperation. There are simple, small grassroots organisations that work on a voluntary basis and are interested in enhancing the everyday lives of individuals and communities in a certain specific neighbourhood in one Global South country. There are also large international NGOs that have clear global strategies, efficient project management practices and thousands of employed persons to implement their global operations. These international NGOs can even influence other major international stakeholders in development cooperation, as well as several aid-donating countries. The objectives of NGOs vary. Besides a general development agenda, they might also have a religious, political, environmental and/or humanitarian focus. The fundraising of NGOs is usually based on donations, crowdsourcing, corporate sponsorship and their own financial assets. Some NGOs are registered as charitable trusts that allow tax exemptions. The annual assistance given by individual NGOs varies from a few hundred to several hundreds of millions of US dollars.

Various types of NGOs are involved in development in different parts of the world. Truly global large NGOs, such as BRAC – the world's largest NGO – are present in almost every one of the Global South countries. Initiated in 1972, it assists over 100 million people annually, employs over 100,000 persons and it also cooperates with private enterprises such as Nike, Inc. (BRAC, 2019). Despite the economic growth in Africa during the early 21st century, poverty is still a very serious everyday phenomenon in the continent. In 17 African countries, the majority of the population lives below the national poverty line, and in only eight countries is the number of people living in poverty less than 25% of the national population (Figure 7.1).

Furthermore, the concentration of immense wealth into a few super-rich individuals and transnational companies has made them important donors for development. For example, large non-profit foundations, such as the Bill and Melinda Gates Foundation and the Chan Zuckerberg Initiative, are also important in international development assistance. They annually donate hundreds of millions USD for development, including to Africa. Large for-profit organisations, such as the American multinational corporation Nike, Inc., also have significant development projects in Africa. This is part of the contemporary corporate responsibility of globally present enterprises (see Chapter 5).

Despite the growth of the aforementioned global actors, a multitude of local grassroots NGOs exist. These are important locally engaged actors that deal with specific local aspects of development. They often consist of committed individuals who share similar ideologies for helping the poor and disadvantaged to develop (see Box 7.2). Some of these NGOs might also have religious or political inclinations, thus supporting their specific agendas while also providing assistance. Others are apolitical and non-religious movements just to help communities in the less developed countries of the Global South.

The development of the Internet has changed NGOs' activities for development. It provides better access to the aid-receiving communities and areas and creates a possibility for interactive information and communications with them. In addition, technological development brings opportunities to solve many issues that were earlier impossible to solve or that would have required huge resources from NGOs. Another reason for change in NGOs and related organisations is the increase of the diversity of stakeholders involved in development cooperation. The international policies and practices of developmental interventions have also evolved. The traditional humanitarian scope is being gradually replaced by more focus

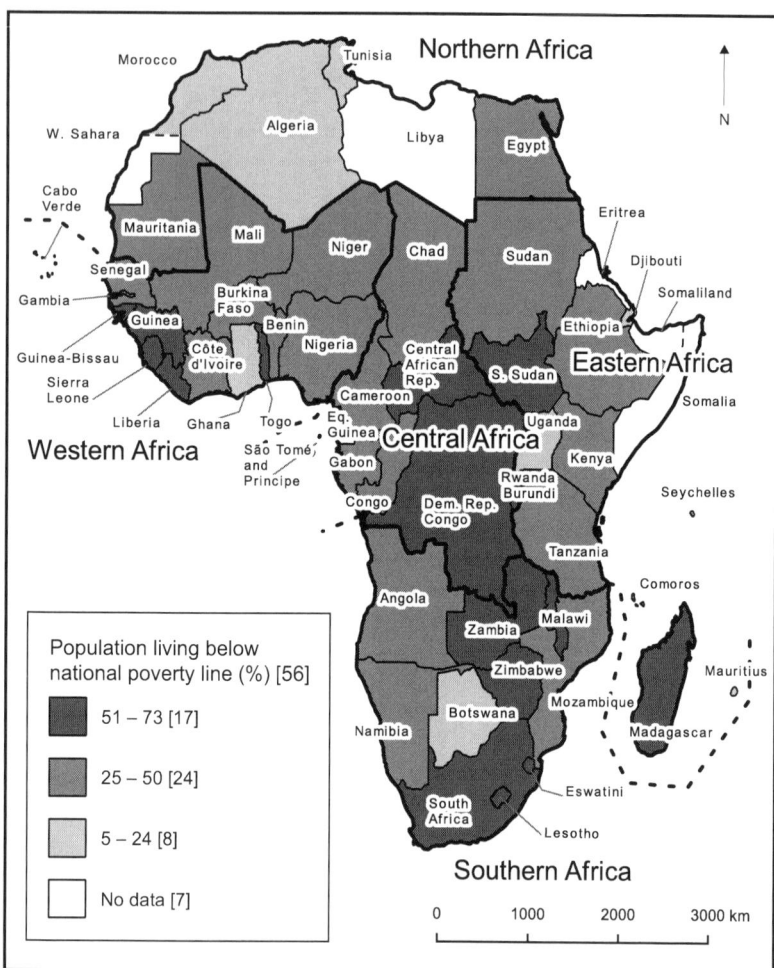

FIGURE 7.1 Poverty in Africa

Source: Modified from UNECA (2018b)

on business, entrepreneurship and innovations (i.e. to help the aid receivers to help themselves).

The market-based approach of development cooperation is changing the role of NGOs in development cooperation. Increasingly, NGOs work in CSPs that are instruments to enhance unilateral actions of governments and civil society (Kolk et al., 2008; Crane, 2010; Vestergaard et al., 2019).

Waddock (1991: 481–482) has defined a CSP as a voluntary collaboration "between actors from organizations in two or more economic sectors in a forum in which they cooperatively attempt to solve a problem or issue of mutual concern that is in some way identified with a public policy agenda item". In many cases, CSPs cover different organisations and institutions that vary between their structural formulation and their objectives. Different partner configurations include various partnerships: public-private, government-NGO, business-NGO and those engaging partners from three or more societal sectors (Selsky & Parker, 2005). Partnership is often enhanced by intergovernmental instruments, such as the SDGs.

From the NGO perspective, the role of a business-NGO partnership is the most relevant in innovation-focussed development cooperation. Information and technology innovations also enable the private sector to expand its markets and make new investments to those locations in Africa where government presence is limited. This has also increased risks of social, political and economic incidents (Idemudia, 2017). In such conditions, NGOs have grown to be mediators between private sector interests and local communities. In particular, NGOs have helped to translate external knowledge into local contexts and indigenous local knowledge to the interests in the outside world. Many committed NGOs have a deep understanding of locally embedded knowledge and local needs and concerns. NGOs can assist the private sector to develop strategies for market and non-market factors and their social, political, environmental and human impacts (Baron, 1995). As a key non-market strategy, a business-NGO partnership can assist corporations to fulfil their social obligations as part of their corporate social responsibility (Bendell et al., 2010; Shah, 2011). Collaborations with NGOs help the private sector to address social challenges and manage complex collective action problems, and thus enhance a firm's corporate social responsibility strategy and make it operationally efficient (Jamali et al., 2011). Furthermore, in weak institutional contexts, partnerships with NGOs are often essential for businesses to secure competitive advantages, control social risks and increase shared value of different stakeholders (Bendell et al., 2010; Idemudia, 2017).

Intensified business-NGO partnerships also have their down sides. For example, close collaboration with market-based actors and profit-seeking collaborations can weaken the reputations and accountability of such NGOs in the eyes of local communities and global donors. Because the private sector and NGOs share fundamentally distinctive values and structures, their relationship can easily be characterised with mistrust and hostility, even when the counterparts share common values and commitments (Rondinelli & London, 2003).

Intergovernmental organisations

Large global intergovernmental organisations have had a major role in the evolution of innovation-focussed development cooperation. Particularly important has been the role of the UN and its sub-organisations, especially UNECA and UNESCO. In addition, the World Bank and the IMF have influenced the structural conditions and development-related policy making in the Global South. As mentioned in Chapters 1 and 4, over the years, the UN has set the general objectives for international development, the most evident with the SDGs. SDG 9 is particularly important for innovation-related development. In the African context, UNECA has for more than 60 years promoted the economic and social development of its African member states. In innovation-related activities, it has convening, think tank and operational functions, especially carrying forward the UN 2030 Agenda with SDGs and the AU Agenda 2068 with its STI development policies. However, as an umbrella organisation, UNECA is somewhat distant from the everyday development of innovation and technology.

STI is also an important objective of UNESCO that provides guidance for revising and developing national STI policies, with special emphasis on Africa. UNESCO's Institute of Statistics collects data and measures countries' progress in STI. Every five years, UNESCO publishes a report that monitors the national status of STI progress. Furthermore, UNESCO fosters linkages between universities and industry with its special programme aiming to translate scientific knowledge to innovations.

The World Bank has had different approaches to development in Africa. While in the 1990s, the main goal was to suggest that African countries invest in primary education instead of higher education, in the 21st century the new suggestion is to invest in STI. In 2000–2013, the World Bank Group invested $18.7 billion USD in various innovation and entrepreneurship initiatives. The World Bank has moved from more straightforward neoliberal policies and macro-level structural adjustments in Africa towards supporting knowledge societies based on innovation, technology development and entrepreneurship. According to the World Bank (2019):

> *Innovation and entrepreneurship are recognized as key building blocks of competitive and dynamic economies. Countries and regions with vibrant innovation and entrepreneurship ecosystems tend to witness higher productivity rates, leading to increased economic growth and more robust job creation, the main pathways through which the poor can escape poverty. As a key driver for firm growth, innovation fosters shared prosperity by stimulating formal employment and increasing wages.*

The World Bank operates at many different levels from individual companies, regions and national innovation systems to transnational innovation politics. Its strategy has four main themes (World Bank, 2019):

(I)   Assessing and diagnosing entrepreneurial ecosystems and identifying their contemporary performance, addressing constraints for success and opportunities for growth of innovative entrepreneurship;

(II) Design of targeted solutions that advocate the innovative capabilities, productivity and growth of firms. This includes developing funding mechanisms, like angel and venture investments, seed funding, and training and support for companies;

(III) Enhancing and strengthening policy design and governance for policy effectiveness. This includes helping governments to review their innovation policies and strategies and public spending on STI, planning and operationalising innovation policies and programmes; and

(IV) Global engagement with key stakeholders of innovation, especially in inclusive and social innovations, as innovation and entrepreneurship enhance the prospects and sustainable growth of the poorest countries.

Foundations

Many transnational companies and super-rich persons have recently established private foundations and charities. Many of these people have backgrounds in technology development and global business, so their favourite theme has been to support innovations, technology development and entrepreneurship. This can also be part of broader corporate social responsibility. As discussed in Chapter 4, supporting innovations is also an effective strategy for these enterprises to create new markets for their goods and services and market them as well.

Significant foundations in innovation-focussed development cooperation are affiliated with ICT. Examples are the Bill & Melinda Gates Foundation originating from the founder of Microsoft, and the Chan Zuckerberg Initiative that was established by Mark Zuckerberg, the founder of Facebook.

The Bill & Melinda Gates Foundation is funding the Grand Challenges Africa Innovation Seed Grants. It is part of the Alliance for Accelerating Excellence in Science in Africa, an initiative that the African Academy of Sciences and the New Partnership for African Development are organising. The fund seeks strategies and solutions to reduce neonatal, maternal and child deaths in Africa through medical innovation. It also seeks creative and effective approaches for communication to get African governments inspired to fund R&D. In the early 2000s, the Bill & Melinda Gates Foundation established with the Rockefeller Foundation the Alliance for a Green Revolution in Africa to help millions of small-scale farmers and their families lift themselves from poverty (Toenniessen et al., 2008). However, some scholars have argued that instead of technological solutions for agricultural improvement, more profound structural reforms in the market and political systems would be needed (Holt-Gimenez et al., 2008).

Andela is a Chan Zuckerberg Initiative-sponsored technology enterprise that offers paid training in software development in various places of Africa. Andela was initiated in 2014 in Nigeria as a start-up company. It started to train software developers in Africa through its boot camp and four-year mentoring programme. During the programme, the participants couple with their trainees and enterprises from the United States that need software developers. In 2015, the Chan Zuckerberg Initiative decided to invest $100 million USD in Andela. In addition to Nigeria, Andela is now also operating in Kenya, Uganda, Rwanda and the United States. Over the years, Andela has hired more than 1,100 remote-based African engineers.

However, other foundations such as the Nike Foundation and Sainsbury Family Charitable Trusts (see Box 6.2) have also been active. Foundations are not only bringing in a huge amount of new money but also new knowledge, networks and facilities to be used in the development cooperation. JA Worldwide is more than 100 years old as an NGO, and among the largest youth employment and entrepreneurship-serving NGOs. Its sub-organisation JA Africa connects entrepreneurs and promotes innovation across the continent. In addition, there are also many specific thematic organisations assisting in development. For example, Water.org has created unique, durable solutions to the water crises in several African countries.

BOX 7.2 INDIGO TRUST STIMULATING TRANSPARENCY, ACCOUNTABILITY AND EMPOWERMENT IN AFRICA

The Indigo Trust aims to create a world of informed, active citizens and accountable, responsive governments that together foster positive change in society. This grant-providing foundation is based in the United Kingdom and it operates in Ghana, Kenya, South Africa and Uganda.

It belongs to the Sainsbury Family Charitable Trusts. It provides small and high-risk grants (£10,000–20,000 GBP) to early-stage projects. The Indigo Trust supports focus on organisations and projects that leverage web and mobile technologies to stimulate innovative approaches to accountability, transparency and citizen empowerment. It also supports co-creational hubs and civic technology communities that use ICT for positive social change.

The accountability and transparency funding concentrates on investigating new ways in which web and mobile technologies can be incorporated into broader agendas, enabling citizens to access the knowledge needed to make informed decisions and keep authorities accountable. This includes access to budget information, parliamentary monitoring, freedom of information, corporate transparency and democratic transparency. The Indigo Trust also cooperates with organisations that apply digital technologies which allow citizens to monitor public service delivery and foster improvements. Another priority is open data groups that apply technology and data to promote informed decision-making for social development.

Local organisations

Small grassroots organisations can have impacts that are bigger than their organisational size in innovation-focussed development cooperation. Most often this is based on their engagement with local and indigenous knowledge and their accumulated social capital with local communities. These non-profit NGOs have often conducted humanitarian activities for a long time in the societies and communities in which they operate, so they have gained the trust of local people. Such NGOs are among the best experts at identifying local needs and ideas for social and inclusive innovations.

Together with the donor countries, NGOs have systematically collected longitudinal data on local development. The data are very valuable for the application of future technologies and artificial intelligence. However, applying such data to for-profit market purposes includes several ethical challenges. The NGOs depend on committed people. Sometimes just one skilled and motivated community member can have a radical impact on local socio-economic development, as the case of Elimu living lab in Sengerema, Tanzania, indicates (see Box 7.3).

BOX 7.3   CREATING SOCIAL ENTERPRISES IN ELIMU LIVING LAB

Elimu is a Swahili word meaning "knowledge" or "education". Elimu Living Lab (for living labs, see Chapter 8) is located in Sengerema, a town of more than 200,000 inhabitants by the Lake Victoria area in northeast Tanzania. The establishment and development of this living lab has been a project of former university student Karol Novat. In 2012, he unintentionally participated in the national capital Dar-es-Salaam's living lab training, organised by TANZICT, the Finland-Tanzania development cooperation programme (see Chapter 4). During the workshop, he became enthusiastic about setting up a living lab in his hometown of Sengerema. He took his personal computer outdoors and began to train people in its usage. Soon almost 200 youths and children gathered around his computer to learn its usage.

The Elimu Living Lab grew rapidly, and during its first two years it impacted more than 1,500 local residents. The success of Elimu was significantly grounded to its founder's personal innovation and entrepreneurial skills. The Elimu Living Lab has established numerous social enterprises and it has become financially self-sustainable. The main source of income has been its printing and computer-design shop located in downtown Sengerema. This has been a beneficial business idea and practice in a town where computer-assisted design, copy machines and knowledge about their usage are scarce. Their main customers are schools and education institutions, the local municipality, private companies and individuals. Later, the Elimu Living Lab created tens of innovative enterprises run by students. For example, they manufactured petroleum jelly, built huge water containers from recycled plastic bottles, opened the first day-care centre of Sengerema and developed a digital management system for Tanzanian schools.

Social entrepreneurship has been an essential part of the Elimu Living Labs curriculum. Its development is based on co-innovation with students, formulation of new business plans and establishing companies. The established companies and trained students with ICT and entrepreneurial skills have created more than 200 jobs. The income from enterprises is used to develop new educational programmes based on the community's needs. The main education program, the Growth Leadership Academy, focusses on entrepreneurial, leadership and ICT skills. The living lab also offers educational programmes for around 200 school dropouts, of whom some are also supported with daily meals and accommodation. In addition, the Elimu Living Lab facilitated innovation space where ICT equipment, computers and wireless Internet are offered to community members. One of the latest initiatives was to establish a new business-oriented vocational training centre.

Rapid growth has also been a challenge for the Elimu Living Lab as the people able to run the living lab's activities efficiently are few and their skills are insufficient for many required activities. Despite the constant capacity-building of the trainer programme, the Elimu Living Lab relies heavily on its founder. To develop ICT-based software and applications further, there is a need to cooperate with higher education institutes and software developers, but these are lacking in the area. In addition, most activities are new and disruptive ways of doing things, and hence receiving necessary permits from the public authorities is challenging. According to the founder, public administrators and municipality officers have been keen to create regulations and disincentives rather than being interactive and cooperative.

Conclusions

Current development cooperation cannot properly operate without NGOs and other non-public and non-private organisations. These organisations, such as charities, have been present in less developed countries for centuries. However, currently their sizes, scopes and practices vary from globally impactful organisations with corporate structures to locally meaningful volunteer groups. Some global organisations, such as the World Bank, IMF, OECD and UN, substantially influence directly and indirectly the development policies and practices in the Global South countries.

Particularly interesting is the relation of NGOs and other non-public and non-private organisations to the official development aid conducted by the countries of the Global North. These organisations do not have to follow official development guidelines and, in fact, many of them also have more than purely developmental goals. The intensifying networks between all stakeholders involved in development aid, assistance and cooperation mean that these organisations are also increasingly in contact with the public and private donors of the Global North.

Another particularity is the rise of super-rich multibillionaires who have started to sponsor less developed countries. In recent years they have become very significant individual donors, for example, in Africa. These individuals, through their charity foundations, can select whatever target they like for donations. Often these novel stakeholders want to contribute to the solving of grand development challenges with the enhancement of technology. Therefore, they often have a clear innovation component in their activities. However, there is also room for community-based development with only small external support. A significant challenge, and also an opportunity, is their dependence on locally motivated, committed and innovative individuals.

- Many kinds of NGOs and other non-public and non-private organisations are involved in development cooperation with different aims, structures and practices.
- Many NGOs and other non-public and non-private organisations involved in development aid and assistance have modified their practices towards cooperation with the private sector and include more economic development on their agenda.
- NGOs and other non-public and non-private organisations can find niches in development cooperation, and they can have potentially substantial and faster development impact.

Discussion questions

- What kind of NGOs and other non-public and non-private organisations are involved in development cooperation in Africa?
- What opportunities and challenges do NGOs and other non-public and non-private organisations bring to the development cooperation in Africa?
- Discuss what role large international foundations can have in supporting development in Africa.

References

Baron, D. (1995). NAFTA and the environment – making the side agreement work. *Arizona Journal of International and Comparative Law* 12, 603–617.
Bendell, J., Collins, E. and Roper, J. (2010). Beyond partnerism: Towards a more expansive research agenda on multi-stakeholder collaboration for responsible business. *Business Strategy and the Environment* 19, 351–355.
BRAC (2019). www.brac.net/. Retrieved June 2019.
Crane, A. (2010). From governance to governance: On blurring boundaries. *Journal of Business Ethics* 94, 17–19.
Edwards, M. (2008). Gates, Google, and the ending of global poverty: Philantrocapitalism and international development. *Brown Journal of World Affairs* 15, 35.

Holt-Gimenez, E., Altieri, M. and Rosset, P. (2008). *Ten Reasons Why the Rockefeller and the Bill and Melinda Gates Foundations' Alliance for Another Green Revolution Will Not Solve the Problems of Poverty and Hunger in the Sub-Saharan Africa.* Agris, FAO, Rome.

Idemudia, U. (2017). Environmental business-NGO partnerships in Nigeria: Issues and prospects. *Business Strategy and the Environment* 26:2, 265–276.

Jamali, D., Yianni, M. and Abdallah, H. (2011). Strategic partnerships, social capital and innovation for social collaboration innovation. *European Ethics: European Review* 20:4, 375–391.

Kolk, A., Van Tulder, R. and Kostwinder, E. (2008). Business and partnerships for development. *European Management Journal* 26:4, 262–273.

Levich, J. (2015). The Gates foundation, ebola and global health imperialism. *American Journal of Economics and Sociology* 74:4, 704–742.

Rondinelli, D. and London, T. (2003). How corporations and environmental groups cooperate: Assessing cross-sector collaboration and collaborations. *Academy of Management Executives* 17:1, 61–76.

Selsky, J. and Parker, B. (2005). Cross-sector partnerships to address social issues: Challenges to theory and practice. *Journal of Management* 31:6, 849–873.

Shah, K. (2011). Organizational legitimacy and strategic bridging ability of green collaborations. *Business Strategy and the Environment* 20, 498–511.

Toenniessen, G., Adesina, A. and DeVries, J. (2008). Building an alliance for a green revolution in Africa. *Annals of the New York Academy of Sciences* 1136:1, 233–242.

UNECA (United Nations Economic Commission for Africa) (2018b). *African Statistical Yearbook 2018.* UNECA, Addis Ababa.

Vestergaard, A., Murphy, L., Morsing, M. and Langevang, T. (2019). Cross-sector partnerships as capitalism's new development agents: Reconceiving impact as empowerment. *Business & Society*, doi: 10.1177/0007650319845327.

Waddock, S. (1991). A typology of social partnership organizations. *Administration & Society* 22:4, 480–515.

World Bank (2019). *Innovation and Entrepreneurship.* www.worldbank.org/en/topic/innovation-entrepreneurship. World Bank, Washington, DC. Retrieved June 2019.

8
LOCAL COMMUNITIES

Introduction

Innovations in the past decades have enhanced the nourishment, health and life expectancy of the African population. They have brought better livelihoods through structural economic changes, improved people's mobility and communication and provided enhanced access to information. Many local communities live nowadays much better than a few decades ago.

However, French economist Thomas Piketty (2014) claims that new technologies and innovations also lead into unsustainable prosperity through their unequal distribution. The geography of innovations and their outputs is very uneven globally (Shearmur et al., 2016). Indian development scholar K.J. Joseph (2014), for example, illustrates how the STI policies of India in the 2010s led to remarkable income and economic growth in the country, but at the same time the income gap widened in the society. Such are the common results in the Global South from top-down innovation policies with a strong focus on economic growth and STI (Chataway et al., 2014). Local communities in the Global South, including in Africa, witness changes in their environments, but they are rarely active and influential stakeholders of societal transformation. The involvement of the world's poor in innovation development has been thin, and they have received hardly any profit sharing from the products and processes built on their knowledge.

Africa still desperately need innovations so that the people could get along better in everyday life. Fresh and clean water, inexpensive and reliable electricity, and better health services are crucial. However, survival cannot be the ultimate development goal in Africa. On the contrary, every African needs to be able to live a meaningful life and contribute to the sustainable development of community and society. The broader shift from dependency to active engagement in development requires pro-poor and pro-African inclusive innovations. For this, African competitive advantages can be found in local assets, such as indigenous knowledge. However, innovations cannot turn exclusively to Africa, but need to be open both to its communities and to the world. Henceforth, the establishment of more inclusive innovation systems that improve the engagement of all communities in Africa is fundamental for local development and global equality and justice (Papaioannou, 2014).

• The use of indigenous knowledge in Africa can be a competitive advantage for long-term sustainability in innovation development.
• Inclusive living labs in Africa are suitable platforms to support systematic, open and inclusive innovation processes.
• A progressive innovation-focussed development cooperation can engage the poor in Africa with innovation creation, systems and policies.

In this chapter, we discuss the emerging modes and practices of inclusive innovations to foster positive transformations and reduce socio-economic inequalities in Africa. After this introduction, we discuss in Section 8.2 the development of local communities and the potential for inclusive innovations. In Section 8.3, we illustrate how indigenous knowledge deriving from different contexts and areas in Africa is utilised for innovation development. Here, we elaborate also the notion of unnovation as a concept for sustainable development by not changing. In Section 8.4, we turn to inclusive living labs as particular platforms to support systematic, open and inclusive innovation processes in Africa. Finally, in Section 8.5, we present the conclusions.

Changing Africa and inclusive innovations

During the last decade, one of the major changes in world development has been the rapid economic growth in Africa. Although the economic growth of the continent has become slower, partially due to a highly uncertain oil market, the growth estimation for 2019–2020 is still around 3% (World Bank, 2019). The main drivers of the growth are the continent's commodities like minerals, oil and tropical agricultural products. Enormous new mineral reserves are frequently discovered. However, the steadiest growth is currently occurring in Ethiopia, Ghana, Côte d'Ivoire, Senegal and Tanzania (World Bank, 2019).

Many challenges remain in Africa. Rapid population growth and urbanisation provide new opportunities for economic growth and innovation, but also pose fundamental challenges, for example, to environmental sustainability, the capacity of the education system, and food security (Rakodi, 2016). The poorest countries in Africa are Somalia, Central African Republic, Democratic Republic of the Congo, Burundi and Liberia. Economic gains are tightly concentrated to a narrow number of locations, sectors, and individuals. Despite the rapid growth, poverty remains high. In 2019, 422 million people in Africa lived in extreme poverty. However, for the first time in the 2010s, the number of poor people started to decline in 2019. It is estimated that this development will continue, and tens of millions of Africans will be lifted from poverty in the 2020s (Figure 8.1).

There is a risk that the grand challenges of Africa and the ongoing commodity export-based development pushes the continent further into dependency, underdevelopment and marginality (Taylor, 2015). Much of the GDP growth derives from a narrow range of extractive commodities and industries without local value addition or contribution to the structural transformation of the economy (Sindzingre, 2013). To respond to these grand challenges, innovation policies in Africa need to support the development of more efficient and inclusive societies (Adebowale et al., 2014). Innovation policy may help in this if it is politically targeted, strategically designed and operationally applied to tackle the grand challenges, and it involves local communities (Altenburg, 2009: 37). However, most innovation policies in African countries fail to create plausible strategies or operational practises to reach these objectives. Therefore, innovations' relationship to inequality and poverty is co-evolving and complex (Cozzens & Kaplinsky, 2009).

According to Papaioannou (2018), the normative considerations of innovation and related processes mean that bottom-up participation, equity, transformative non-hierarchical models, or indigenous knowledge are very seldom appreciated. The involvement of local communities requires a turn toward inclusive, transformative, and participatory innovations. These are based on, for example, user-centred, non-technological, and service-oriented incremental processes (Chataway et al., 2014). Although innovation is often presented as apolitical, it always contains normative implications of what knowledge, whose interests, and which outcomes are significant (Bryden et al., 2017). Innovation needs to be developed and evaluated as a social process.

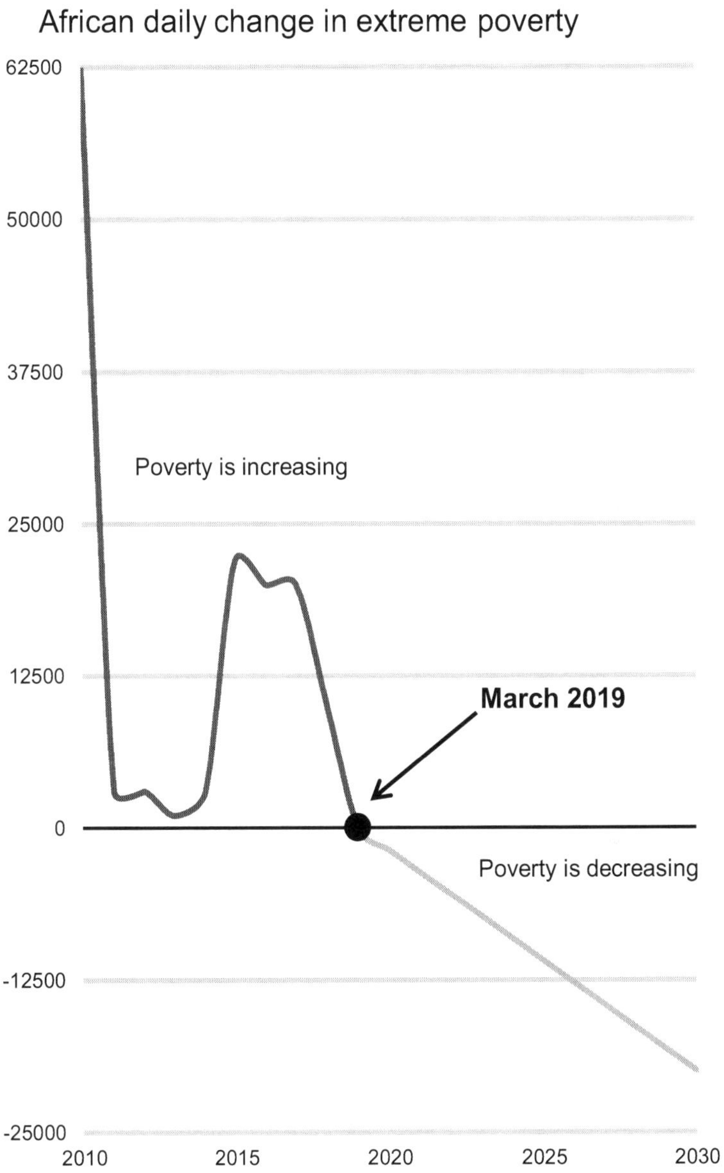

African daily change in extreme poverty

62500

50000

37500

Poverty is increasing

25000

12500

March 2019

0

Poverty is decreasing

-12500

-25000

2010   2015   2020   2025   2030

FIGURE 8.1   Development of poverty in Africa, 2010–2030

Source: Modified from Kharas et al. (2019)

BOX 8.1   INCLUSIVE INNOVATIONS

In the early 21st century, many scholars demonstrated how technological innovations failed to acknowledge the needs of the global poor. As an alternative, scholars identified and developed better ways to include pro-poor products, services, and organisational innovations (Arocena & Sutz, 2003; Muchie, 2004; Lundvall et al., 2009; Arocena & Sutz, 2014; Chataway et al., 2014; Papaioannou, 2014; Bryden et al., 2017; Planes-Satorra & Paunov, 2017; Schillo & Robinson, 2017; Papaioannou, 2018; Jímenez, 2019). These include various concepts, such as frugal innovations, bottom of the pyramid (BoP) innovations, pro-poor and from-the-poor innovations, below the radar innovations, and responsible innovations (see Chapter 2). The OECD branded these different concepts as inclusive innovations (see Paunov, 2013). These can have significant positive outcomes to the welfare of the local poor in developing countries. Inclusiveness is necessary, as innovations are expected to contribute substantially to solving the grand global and continental challenges (Kallerud et al., 2013) and to move toward transformative outcomes (see Chapter 2).

Inclusiveness is generally described as the opposite of social exclusion (Papaioannou, 2018). Traditional innovation policies usually focus on large private and public organisations, addressing the interests of middle- and higher-income customers (Trojer et al., 2014). Foster and Heeks (2013: 4) identified five aspects why such innovation processes have failed to be inclusive. First, formal innovations focus insufficiently on the poor. Second, informal actors are delinked from innovation systems. Third, innovations aimed to support peripheral areas are often weakly adaptive. Fourth, low-income innovation users lack the capability to use these effectively. Fifth, supporting policies and contexts for innovation creation are weak or absent.

Inclusive innovation is about open (Chesbrough, 2003; Chesbrough & Appleyard, 2007) and democratic participation in innovation processes (Von Hippel & Katz, 2002; Von Hippel, 2015) and novel social configurations to organise more equal innovation development (Mensink et al., 2010). Most inclusive innovations promoted by the international development community are based on the objectives of the SDGs (see Chapter 3). Inclusive innovations include people at the bottom of the global income pyramid (i.e. the BoPs). In many innovation processes, they are seen merely as end users of the innovation outputs. They are rarely considered innovation developers improving their own well-being. In the Global South, innovation developers often originate from outside of this region. Lacking an in-depth understanding of local socio-economic contexts, they are unable to identify the innovation needs and opportunities of the global poor (Kaplinsky, 2011). Even if they recognise the structure of technology components and value changes, it is difficult for them to utilise these opportunities. Nevertheless, inclusive innovation processes emphasise not only economic but also social and environmental outcomes. For inclusion, the recognition of the DUI mode of achieving innovation is important (Hooli et al., 2019).

Agriculture is a fast-developing field for inclusive innovations and technologies in Africa (Nakasone et al., 2014). The continent's rapid population growth and urbanisation require more productive modes of agriculture, including in urban areas, for example, in forms of guerrilla gardening. According to the African Development Bank (2018), around 60% of the world's uncultivated, arable land is in Africa. There are high expectations that Africa could become the global food basket as the next frontier of the Green Revolution.

Smallholder farming is still today the most common livelihood among the vast majority of the rural population in Africa, where more than two-thirds are employed in agriculture. According to Hounkonnou et al. (2012), the increase in the productivity of African smallholders is key to global food security. A lack of enabling institutions explains much variance in the quantity and quality of smallholder output in Sub-Saharan Africa, where technology-driven productivity growth has largely failed. Therefore, besides the large-scale modernisation of agriculture, one should also pay attention to informal experimentation and smaller local innovations that work in the context of small-scale agriculture (Scheuermaier et al., 2004). Small-scale agriculture is a key sector, where inclusive innovations can have a substantial positive impact.

In the end of the 2010s, new agricultural technologies in Africa have gained a large interest among donors, venture capital investors, and new start-up enterprises. International donors have supported dozens of agricultural innovation projects in almost every corner of Africa. In the following, we highlight some of them. Broader research-based agriculture-focussed programmes aim to generate a positive change by investing in innovation development. For example, the Europe-Africa Research and Innovation created the programme LEAP-Agri as a grand initiative related to the food and nutrition security and sustainable agriculture in Africa. In 2016–2021, the project had 30 partners, including 24 ministries and funding agencies from 18 African and European countries. With a budget of approximately €23 million EUR, the initiative funded 27 research and innovation projects involving 250 African and European teams from 20 countries. Agricultural innovations, technologies and participatory processes were highlighted. The cooperation was an important aspect, so each project involved at least two teams from African countries and two teams from European countries (see www.leap-agri.com). Other innovations for agriculture require more advanced technological knowledge and capacity building. For example, SIDA implemented its BioInnovate program in 2009–2021 to build the capacity of innovators, scientists, researchers, and entrepreneurs to discover new bio-based innovations, biological research technologies, and research ideas to develop markets that improve the livelihood of local communities and farmers (BioInnovate, 2019; see Chapter 4). Innovations can also originate from introducing technological applications into a new context. For example, the South African Aerobotics and WeFly from Côte d'Ivoire developed as drone software companies with a data-analysis platform using artificial intelligence to allow farmers to optimise their agricultural output.

Such promotion of agriculture-related technology development in Africa is not only from the Global North, but also from other countries. For example, the huge China-Africa Development Fund (CADFund), established in 2007 and solely funded by the Chinese government policy bank China Development Bank, invests in Chinese enterprises with economic and trade activities and investments in African businesses and projects. By 2018, CADFund's investment decisions reached approximately $5 billion USD for nearly 100 projects in 36 African countries. In agriculture, the main aim has been modernisation, with the improvement of African agricultural capacity and productivity primarily through experience-sharing and technology transfer. Another aim is to encourage profitable Chinese agricultural investments in Africa and bring Chinese experts to African countries (Sun, 2018). According to estimations, Chinese companies are using approximately 5% of arable lands in Central, East and West Africa. With the support of the CADFund, Chinese agricultural companies established large farms in Africa and at least 25 agricultural technology demonstration centres. These centres integrate with local agricultural industries and create footholds for Chinese companies in the new markets, for example, in Tanzania, Mozambique and Rwanda.

To create new solutions to global food production, the German Federal Ministry for Economic Cooperation and Development (BMZ) supports food production and agricultural innovations in 14 partner countries with locally adaptive ways. The BMZ funded the establishment of green innovation centers with €278 million EUR. The objective is to apply innovative techniques, methods and forms of organisations to develop local agriculture. This includes the use of high-quality seeds and fertilisers, new technical innovations like modern harvest and farming techniques, and development for new cooling systems for storage. The centres are based on new forms of cooperation such as producer groups of farmers, capacity-building and education, financial loans and other arrangements focussing specially on women and youths. According to the BMZ's estimations, the innovation centres' products and services will improve the living conditions of seven million people in Benin, Burkina Faso, Cameroon, Ethiopia, Ghana, India, Kenya, Malawi, Mali, Mozambique, Nigeria, Togo, Tunisia and Zambia.

Numerous new agricultural technology start-up companies emerge all around Africa with accelerating speed. Many of their innovations derive from users and are based on the collaboration between users and technology producers for improved products. Often such African innovation projects for agriculture are small and have simple applications that can have large impacts on the lives of farmers. Many applications provide farmers with information about the prices of agricultural goods in the market and give practical advice for agriculture-related activities. Three examples are iCow and Twiga Food in Kenya (see Chapter 3) and e-Soko in Rwanda. In 2010, the government of Rwanda established the eRwanda project. Within the project, the World Bank's International Development Association funded the ICT for Development Programme to develop a digital platform to enhance the well-being of small-scale agricultural producers. The award winning e-Soko sends up-to-date market prices of agricultural goods to the users' mobile phones. More than half of the application users are women (Republic of Rwanda, 2019). Twiga Food is a mobile-based, cashless, business-to-business supply platform enabling farmers to interact with markets. It connects smallholder farmers in rural Kenya to informal retail vendors in cities. As a result, the vendors can order and sell fresh produce from farmers across Kenya at competitive prices. Innovation platforms can also be sources for crowdfunding for small-scale agricultural improvements (Twiga, 2019). For example, Farmcrowdy from Nigeria, Ari Farm from Somalia and Livestock Wealth from South Africa improve the financial situation of many smallholder farmers.

Despite the introduction of many helpful technologies, as discussed previously, most rural areas in Africa still suffer from poverty. Various barriers hinder the growth of efficiency in agriculture. These are, for example, inadequate physical infrastructure (lack of roads, electricity and irrigation), poor access to the markets, inconsistent supply chains, lack of human capital, and illiteracy to new technologies. The use of smart phones has helped farmers and other people in agricultural activities enhance their livelihoods. However, the relative number of smart phones and access to the Internet are still remarkably lower in rural Africa compared with cities. In many rural areas, farmers are still not able to receive accurate weather forecasts, knowledge about proper use of fertilisers and soil management or accurate information about the changing pricing in the markets for their agricultural goods. In their comprehensive analysis of responsible low-technology innovations in the Global South, Hartley et al. (2019) concluded that low-technology innovators have challenges to engage users in the innovation development processes, despite there being a need for it and it benefitting the innovation end-users.

Inclusive innovations have addressed the poor communities in the Global South. However, Jímenez (2019: 46) argues that inclusive innovations should especially focus on those marginalised actors who do not feature in the common innovation discourses and narratives. For example, female entrepreneurs only seldom appear in innovation strategies despite the fact that there are more women than men entrepreneurs in Africa, and Ghana has the highest globally relative amount (46.4%) of female entrepreneurs (Master Card, 2018). Nevertheless, the average income level of African women entrepreneurs is significantly lower than that of men. Many women entrepreneurs live below the poverty line, as most work in the informal sector's small businesses with very limited attention to innovation. The lack of financial resources and growth-oriented entrepreneurial strategies create constraints, especially for women, as a study about Ghanaian women entrepreneurs in the tourism sector indicates (Ali, 2018). Furthermore, a more general gender inequality is still a serious challenge in most parts of the continent.

Women have long been a target group for donor development agendas in Africa. Additionally, in innovation activities and entrepreneurships, many donors especially focus on the empowerment of women. For example, in 2018, the German Federal Ministry of Education and Research (BMBF) updated its Africa Strategy that has a total budget of €300 million EUR. Now, the strategy emphasises women's participation in higher education, innovation and research. One crosscutting theme is to promote women scientists transferring their research results to innovations in industry, society and policymaking (BMBF, 2018). In addition to various projects, donors sponsor annual summits, conferences, forums and awards fostering women's entrepreneurship and innovation.

Another more local initiative is the provision of sustainable solar energy to rural African villages. Innovation Africa Israel Ltd. provides solar energy for those locations and situations where it empowers women in particular.

Women use solar energy, for example, to pump clean water to villages and irrigation, and provide maternity clinics and schools with electricity. These innovations bring much-needed technology solutions to rural villages. However, they are inclusive only as end products, and local communities are weakly included in the innovation process. For example, the solar energy project uses exclusive Israeli technology monitored and maintained from a distance.

Indigenous knowledge in innovation development in Africa

Indigenous knowledge is one possibility for inclusive innovation processes and contextualised innovation policies in Africa. Indigenous knowledge is a local context-specific knowledge that is accumulated over time; unique to a particular society, culture, or local community; and embedded in its activities (Sillitoe & Marzano, 2009). Indigenous knowledge is most often diffused orally or through demonstration and imitation and learned through repetition (Subba, 2006). Therefore, the DUI mode of learning is important for its development. However, indigenous knowledge is always influenced by external knowledge. It may be supported by new external knowledge that is then adjusted to local socio-economic context, or it can contribute to innovations external to the origin of indigenous knowledge (Weichselgartner & Kasperson, 2010).

Indigenous knowledge is important for inclusive innovation processes. It concerns many kinds of products, services, business models, institutions, and supply chains (George et al., 2012). It is the most accessible and applicable knowledge for poor, rural African communities' daily livelihoods (Hagar, 2003; Domfeh, 2007). Concrete innovations resulting from indigenous knowledge vary as much as their usage. Examples are traditional plants for new medicines, tourism activities based on local traditional cultures, increases in resilient practices against negative impacts of climate change, and certain local social transformations such as eradication of female circumcisions. The understanding of indigenous knowledge is important in Africa to comprehending the local communities' ways of interpreting, observing and discussing local issues and world issues. However, more than any single innovative product or service, indigenous knowledge can facilitate participatory processes in innovation development in Africa, and embed local knowledge into broader innovation processes.

In recent years, indigenous knowledge has become a strategic direction of innovation policies in many African countries, such as Botswana, Ghana,

Namibia, Malawi, South Africa and Tanzania (Nfila & Jain, 2011; Jauhiainen & Hooli, 2017). Also, large development agencies like the World Bank (2010) acknowledge that local people at economic and social margins are capable of utilising indigenous knowledge for innovation development. Furthermore, local universities have undertaken indigenous knowledge for their research, teaching and innovation curriculum. As one example, the University of Botswana created the Centre for Scientific Research, Indigenous Knowledge and Innovation to connect indigenous knowledge with innovations. However, most strategies and innovation policies in Africa regard indigenous knowledge rather simplistically. Its incorporation in the innovation process is considered to be a very straightforward practice, in which its instrumental value can be enhanced in the innovation processes and afterwards incorporated into products and services new to local, national and international markets (see Box 8.2).

BOX 8.2 BENEFIT SHARING OF INDIGENOUS INNOVATIONS: THE SAN-HOODIA CASE

Indigenous knowledge has been cleverly and properly used in many contemporary innovation processes, for example, to develop new foods, cosmetics and medicines. However, its application is not always without challenges. One example is the so-called San-Hoodia case. It illustrates the complexity of commercialisation of indigenous knowledge to commercial innovations (see World Health Organization, 2006: 345; Wyndberg et al., 2009).

Hoodia (*Hoodia gordonii*) is a cactus plant that grows in the Kalahari Desert in Botswana, Namibia, and South Africa. San, the oldest inhabitants and indigenous hunter-gatherer groups living in Southern Africa, have used hoodia for ages. The San people use hoodia to treat high blood pressure, gout and diabetes. Most importantly, it is used as an appetite suppressant. It has been used for centuries to prevent thirst and hunger, which is vital for survival in the desert areas.

South Africa's premier scientific research and development organisation, the Council for Scientific and Industrial Research (CSIR), discovered the potential value of hoodia as an anti-obesity product. CSIR patented the active ingredient of the plant. The patenting was done without acknowledging the San people's claims to hoodia or that it derived from their indigenous knowledge. The patent was licensed to a British private company that later sold it to the multinational pharmaceutical enterprise Pfizer, who further sold it to the multinational food industry company Unilever. The aim was to create a huge international economic profit out of the final product.

The plans to capitalise hoodia initially created national and, later, international public pressure. Under marginalisation, colonialism and genocides, about 100,000 San still remain but most live in poverty. Nevertheless, they have maintained indigenous and traditional ways of life and also use the hoodia plant. In 2003, the South African San Council managed to negotiate a benefit-sharing agreement with Unilever that would have ensured the San people received a fair revenue from the commercialisation of hoodia products. In 2006, the agreement was signed with the San and the South African Hoodia Growers (Pty) Ltd. In 2008, after spending more than €20 million EUR for the product development and testing, Unilever withdrew from the commercialisation of hoodia as an anti-obesity succulent. They mentioned the cost of clinical research and product marketing as the main reasons.

Similar challenges also exist with other indigenous plants across Africa. Another example is the protein-rich Marama bean (*Tylosema esculentum*) that the Khoisan people utilise in Namibia. The plant grows uncultivated in the wild. Nevertheless, there are already many registered patents regarding Marama products and processes. The complex benefit-sharing concerning Marama is settled with the patent holders, the indigenous Khoisan people, and the government of Namibia (Percy et al., 2010).

There are also several challenges in the utilisation of indigenous knowledge in innovation-related policies. As discussed previously (Box 8.2), it is not easy to fit traditional indigenous knowledge into contemporary commercial innovations, and such action may harm those whose lives have been supported for generations with indigenous knowledge. In addition, there is a risk of treating indigenous knowledge as a universal common property shared by all individuals in specific regions or countries. For example, in the national development strategies of Namibia, indigenous knowledge is clearly mentioned as a development asset (Jauhiainen & Hooli, 2017). However, the strategy treats it as a general homogenous knowledge that is disseminated equally across the country. This is despite Namibia consisting of more than 10 ethnic groups, of whom many have their own traditions and cultures that differ substantially from cultures and traditions of other ethnic groups. Therefore, some individuals' lives are more immersed in indigenous knowledge than others' lives. Indigenous knowledge is, by its character, always tied to local practices that make it difficult to generalise it and scale it up. Deeply rooted indigenous knowledge processes may also create local cultural and cognitive path-dependencies. These hinder the immersion of external innovations, but also provide possibilities for a very long-term sustainable development, at least in particular contexts (see Box 8.3). Indigenous knowledge has been, so far, seldom acknowledged in the donor countries' development policies or their related innovation strategies toward Africa.

To incorporate indigenous knowledge into an innovation system, the knowledge providers and exploiters of these localities, in which indigenous knowledge resides, need to be acknowledged as well as the local context. Modes and platforms to combine different knowledge bases can be different. Financial resources are needed to enhance knowledge development and learning by combining the past with the present, the local with the non-local, and improving the absorptive capacity at local levels in Africa (Ndabeni, 2016). Inclusive living labs is one such platform, as we illustrate in the next section.

BOX 8.3   UNNOVATION IN AFRICA

Despite change and transformation being key elements in innovations and innovation processes, unnovation (i.e. the ability of not changing while being sustainable) can also be an innovation. The reliance on indigenous knowledge and stable local practices may seem a backward approach in the rapidly changing society. However, at least in particular circumstances, the reliance on indigenous knowledge and the rejection of modern technologies can be a possibility for sustainable developments, livelihoods and cultures.

The Hadza people consist of approximately 1,000 hunter-gatherers who live in many small groups close to the Eyasi salt lake in northern Tanzania. They once lived in an area of more than 10,000 square kilometres. According to genetic and linguistic research, the Hadza have lived in the area for more than 10,000 years. Nowadays, their area has been reduced to a quarter of the original size, mostly due to the growth of farming and livestock expansion by the population initially living outside of this area.

The lifestyles of the Hadza with nomadic camps have not changed substantially over a long time. Their food consists of unprocessed animals and plants, including eggs, vegetables, fruits, nuts and seeds, and their regular physical exercise keeps them healthy and fit. The so-called Paleo diets have taken inspiration from this and other tribes, whose dietary regimes are like it was before modern agriculture appeared several thousands of years ago. Today, the Hadza still do not have electricity, fixed or mobile phones, the Internet, running water or sewage systems. Only a few of them are able to read or write despite the authorities' offers to their children for schooling possibilities. They even try to oblige the Hadza children to receive formal school education.

This glimpse into the past is by no means a suggestion that instead of keeping up with new technologies and higher education, the African people should turn to the past. The Hadza should not be seen as an embarrassment for a modernising nation or a tourism attraction of "the real Africa". Instead, their case shows how indigenous knowledge created thousands of years ago can still be a vital source for living, though in rather particular circumstances and isolated locations. In contemporary innovation creation, there are issues to be learned from indigenous knowledge on how to achieve sustainable development for a time span that extends to hundreds of generations.

Inclusive living labs – organising communities' innovation processes

Living labs have become a common tool around the world to sense, test, and validate new products, services, and solutions in real life contexts. In the Global North, living labs often deal with high-technology development in user-centred innovation ecosystems that are open for interested individuals and stakeholders. In Africa, one finds such technologically advanced living labs in specific innovation districts, usually connected to the most prominent universities and their technology parks (see Chapter 6). As an organisational structure, a living lab is also a process innovation that is becoming widely diffused. In Africa, living labs can be found in, at least, Angola, Algeria, Burkina Faso, Burundi, Botswana, Cameroon, Egypt, Ethiopia, Eritrea, Ghana, Kenya, Lesotho, Malawi, Mauritius, Mozambique, Namibia, Rwanda, Sierra Leone, Senegal, South Africa, Tanzania, Tunisia, Uganda and Zambia.

However, in Africa, living labs can also be a platform to support open inclusive innovation processes (Hooli et al., 2016). The concept of an inclusive living lab in African contexts shifts the community development focus toward the reinforcement of transformative actions from below. It promotes learning, capacity-building, and empowerment in local communities. The emphasis of the inclusive living labs is on co-creational innovation activities together with communities. Different community members interact in this platform to solve and turn the locally experienced common challenges into innovation potential (Hooli et al., 2016). In Africa, inclusive living labs are usually rather modest venues, offering access to electricity, basic ICT facilities, and the Internet. They aim to become self-sustainable, though their establishment is sometimes supported by the community. To fund their activities, inclusive living labs run several small business activities, such as ICT courses and services for companies and the public sector, cafés, copy shops and selling local agricultural goods.

The operational principles of inclusive living labs in Africa resemble those of technology and other co-creation hubs (see Chapters 5 and 6). Technology hubs gather local experts with technology development skills and/or university education backgrounds and connect them to external knowledge sources and business environments. Instead, inclusive living labs function at grassroots levels and involve community members who have low educational levels or are without any formal education backgrounds. Inclusive living labs aim to enhance individuals' and communities' basic technological capabilities, social media skills, entrepreneurial mindsets and problem-solving capabilities (Buitendag et al., 2012; Coetzee et al., 2012). The participation in the activities does not necessarily require previous ICT skills or entrepreneurial experiences. In inclusive living labs, young adults are the most common target group. For example, RLabs in South Africa, as well as several other inclusive living labs, mainly focus on youths with special backgrounds and needs, such as the unemployed, school dropouts, the professionally less qualified and youth with legal or drug problems. Most training and capacity-building in inclusive living labs are conducted by volunteer senior members and peer groups.

Inclusive living labs in Africa are also mechanisms to connect local knowledge and doing-using-interacting (DUI) practices to broader innovation system dynamics. As hubs, they connect various innovation stakeholders at different spatial scales and become intermediate platforms in the knowledge exchange between different entities, such as communities, private enterprises, universities, public authorities and NGOs (Leminen & Westerlund, 2012). Inclusive living labs help to intertwine the innovation creation processes of the knowledge from local communities, as well as exogenous knowledge pools of external experts (Buitendag et al., 2012; Bathelt & Cohendet, 2014). Knowledge creation processes in these inclusive living labs are based on interpreting, applying and refining knowledge created elsewhere which is now being applied to new socio-economic local contexts. The combining of local with external knowledge identifies the needs of local communities and later develops, uses, and markets need-based innovations. The enhanced access to external knowledge via the Internet, social media and ICTs remarkably improves the local communities' entries to the solutions developed in other places. When these solutions are merged with local knowledge, indigenous in some cases, it is an innovation that is not new to the world but new in that specific context.

The access to local knowledge and the interest of their stakeholders in innovation activities makes living labs an interesting partner for foreign or multinational companies, universities and technology companies developing need-based and responsible innovations. This can mean everything between simple data collection and functioning as an inexpensive test bed in communities to scaling up and digitalising innovations developed in the living labs.

The prevailing discussion of innovations in Africa, as elsewhere, is often about high-technology equipment, software applications, and widely applied organisational structures with carefully promoted brands. Small community-level innovations are generally neglected in academic and policy discussions, although they are the most common and widely diffused innovations in Africa. These innovations are developed in local communities and are based on simple knowledge practices and acute everyday challenges there. For example, the establishment of the first kindergarten in the village of Segrenema, Tanzania, created employment, provided early childhood education, and allowed mothers to work. In Harare, Zimbabwe, the rainwater containers are manufactured from recycled plastic bottles and clay. The improvement of the waste management system in Mali increased employment and well-being in communities, though it did not receive attention outside of these communities. As these examples show, there is room for everyday innovations at many levels in Africa. Inclusive living labs have become one of the most successful inclusive models to organise innovation activities and acknowledge indigenous knowledge in local African communities.

Conclusions

Innovations have become key opportunities for the development of local communities in Africa, likewise in the Global North. However, the differences of the socio-economic contexts inside Africa, and especially between the continent and the more developed Global North, mean that the goals and approaches need to be different in Africa. Therefore, the outcomes of innovation for development differ in Africa. Also, the situations in local communities vary, so contextualisation of innovation-related activities is needed.

There are hundreds of millions of people who live below the poverty line in Africa. Most are little or not at all involved in innovation development and the profits originating from them. These people's roles are merely to be the end consumers of innovations, so their participation and needs are not necessarily addressed in innovation development. Therefore, inclusive innovation policies and practices are of particular importance in Africa. These make local people active stakeholders in their own futures, help formulate more consistent policies and practices to reduce poverty and socio-economic inequalities, and foster positive transformations.

Various attempts have tried to make innovations more inclusive in Africa. Global donors see the potential of development cooperation focussing on innovations, so they have started to support local communities. Such cooperation is a possibility to engage the poor and marginalised in the innovation creations, systems, and policies, but only when the practices are tuned through participatory approaches to meet the challenges and opportunities of the poor. The global donors have supported, for example, agriculture and the empowerment of women, which are both crucial aspects in the socio-economic development of Africa. It is significant to identify local practical challenges and needs, combine and access local and external knowledge pools, and mobilise and connect local and external innovation actors and resources to tackle these challenges.

Inclusive living labs are one tool to enhance local communities' skills and competences to be active in systematic, open and inclusive development and innovation processes. A particular asset for innovation development is indigenous knowledge that derives from different contexts and areas in Africa and can source innovation processes better in different local contexts to achieve long-term sustainability.

- The contexts of local communities being engaged with innovations differ between the Global North and Africa, and also inside the continent, so related policies and practices need to be diverse.
- The Global North donors are interested in development cooperation and supporting innovations for development in Africa, but often they do not take enough notice of the complex socio-economic situations in Africa and the participation of the people.

- Inclusive innovations generated and connected to local contexts and indigenous knowledge can provide competitive advantages in Africa and help the marginalised poor to achieve long-term sustainability through innovation for development: inclusive living labs are an important tool.

Discussion questions

- What is an inclusive innovation, and what beneficial impacts do inclusive innovations bring to Africa?
- What opportunities and challenges do local communities in Africa have to develop and enjoy innovations?
- Discuss how the creation and distribution of inclusive innovations can be supported among local communities in Africa.

References

Adebowale, B., Diyamett, B., Lema, R. and Oyelaran-Oyeyinka, O. (2014). Introduction. *African Journal of Science, Technology, Innovation and Development* 6:5, v–xi.

African Development Bank (2018). *Agri-Tech Can Turn African Savannah into Global Food Basket*. www.afdb.org/en/news-and-events/agri-tech-can-turn-african-savannah-into-global-food-basket-african-development-bank-18609/. Retrieved May 2019.

Ali, R. (2018). Determinants of female entrepreneurs growth intentions: A case of female-owned small businesses in Ghana's tourism sector. *Journal of Small Business and Enterprise Development* 25:3, 387–404.

Altenburg, T. (2009). Building inclusive innovation systems in developing countries: Challenges for IS research. In Lundvall, B., Joseph, K., Chaminade, C. and Vang, J. (eds) *Handbook of Innovation and Developing Countries: Building Domestic Capabilities in a Global Setting*, 33–56, Edward Elgar, Cheltenham.

Arocena, R. and Sutz, J. (2003). Knowledge, innovation and learning: Systems and policies in the north and in the south. In *Systems of Innovation and Development: Evidence from Brazil*, 291–310. Edward Elgar, Cheltenham.

Arocena, R. and Sutz, J. (2014). Innovation and democratisation of knowledge as a contribution to inclusive development. In Dutrénit, G. and Sutz, J. (eds) *National Innovation Systems, Social Inclusion and Development. The Latin American Experience*, 15–33. Edward Elgar, Cheltenham.

Bathelt, H. and Cohendet, P. (2014). The creation of knowledge. Local building, global accessing and economic development – toward an agenda. *Journal of Economic Geography* 14:5, 869–882.

BioInnovate Africa (2019). *About Us*. https://bioinnovate-africa.org/about-us/. Retrieved June 2019.

BMBF (German Federal Ministry of Education and Research) (2018). *The Africa Strategy of the BMBF*. www.bmbf.de/. Retrieved June 2019.

Bryden, J., Gezelius, S., Refsgaard, K. and Sutz, J. (2017). Inclusive innovation in the bioeconomy: Concepts and directions for research. *Innovation and Development* 7:1, 1–16.

Buitendag, A.,Van Der Valt, J., Malebane, T. and de Jager, L. (2012). Addressing knowledge support services as part of a living lab environment. *Issues in Informing Science and Information Technology* 9, 221–241.

Chataway, J., Hanlin, R. and Kaplinsky, R. (2014). Inclusive innovation: An architecture for policy development. *Innovation and Development* 4:1, 33–54.

Chesbrough, H. (2003). *Open Innovation: The New Imperative for Creating and Profiting from Technology*. Harvard Business School Press, Boston, MA.

Chesbrough, H. and Appleyard, M. (2007). Open innovation and strategy. *California Management Review* 50:1, 57–76.

Coetzee, H., Du Toit, I. M. and Herselman, M. (2012). Living Labs in South Africa: An analysis based on five case studies. *The Electronic Journal for Virtual Organizations and Networks*, 14. http://hdl.handle.net/10204/6082. Retrieved September 2019.

Cozzens, S. and Kaplinsky, R. (2009). Innovation, poverty and inequality: Cause, coincidence, or co-evolution? In Lundvall, B., Joseph, K., Chaminade, C. and Vang, J. (eds) *Handbook of Innovation and Developing Countries: Building Domestic Capabilities in a Global Setting*, 57–82. Edward Elgar, Cheltenham.

Domfeh, K. (2007). Indigenous knowledge systems and the need for policy and institutional reforms. *Tribes and Tribals* 1, 41–52.

Foster, C. and Heeks, R. (2013). Conceptualising inclusive innovation: Modifying systems of innovation frameworks to understand diffusion of new technology to low-income consumers. *European Journal of Development Research* 25:3, 333–355.

George, G., McGahan, A. and Prabhu, J. (2012). Innovation for inclusive growth: Towards a theoretical framework and a research agenda. *Journal of Management Studies* 49:4, 661–683.

Hagar, C. (2003). Sharing indigenous knowledge: To share or not to share? That is the question. In *Proceedings of CAIS/ACSI 2003 Conference*, Ontario, Canada, May 30–June 1.

Hartley, S., McLeod, C., Clifford, M., Jewitt, S. and Ray, C. (2019). A retrospective analysis of responsible innovation for low-technology innovation in the Global South. *Journal of Responsible Innovation* 6:2, 143–162.

Hooli, L., Jauhiainen, J., Järvi, A., Nkonoki, E., Taajamaa, V. and Käyhkö, N. (2019). Contextualising innovation in Africa: Knowledge modes and actors in local innovation development. In *IST-Africa Week Conference (IST-Africa) Proceedings*, Nairobi, Kenya, 2019.

Hooli, L., Jauhiainen, J. and Lähde, K. (2016). Living labs and knowledge creation in developing countries: Living labs as a tool for socio-economic resilience in Tanzania. *African Journal of Science, Technology, Innovation and Development* 8:1, 61–70.

Hounkonnou, D., Kossou, D., Kuyper, T. and Leeuwis, C. (2012). An innovation systems approach to institutional change: Smallholder development in West Africa. *Agricultural Systems* 108, 74–83.

Jauhiainen, J. and Hooli, L. (2017). Indigenous knowledge and developing countries' innovation systems. The case of Namibia. *International Journal of Innovation Studies* 1:1, 89–106.

Jímenez, A. (2019). Inclusive innovation from the lenses of situated agency: Insights from innovation hubs in the UK and Zambia. *Innovation and Development* 9:1, 41–64.

Joseph, K. (2014). Exploring exclusion in innovation systems: Case of plantation agriculture in India. *Innovation and Development* 4:1, 73–90.

Kallerud, E., Amanitidou, E., Upham, P., Nieminen, M., Klitkou, A., Olsen, D., Toivanen, M., Lima, M., Oksanen, J. and Scordato, L. (2013). Dimensions of research and innovation policies to address grand and global challenges. *NIFU Working Paper* 2013:13.

Kaplinsky, R. (2011). Bottom of the pyramid innovation and pro-poor growth. In Dutz, M., Kuznetsov, Y., Lasagabaster, E. and Pilat, D. (eds) *Making Innovation Policy Work: Learning from Experimentation*, 49–70. OECD, Paris.

Kharas, H., Hamel, K., Hofer, M. and Tong, B. (2019). Global poverty reduction has slowed down – again. *Brookings*, May 23.

Leminen, S. and Westerlund, M. (2012). Towards innovation in Living Labs networks. *International Journal of Product Development* 17:1–2, 43–59.

Lundvall, B., Joseph, K., Chaminade, C. and Vang, J. (eds) (2009). *Handbook of Innovation and Developing Countries: Building Domestic Capabilities in a Global Setting.* Edward Elgar, Cheltenham.

Master Card (2018). *Master Card Index of Women Entrepreneurs.* newsroom.mastercard.com/wp-content/uploads/2018/03/MIWE_2018_Final_Report.pdf/. Retrieved June 2019.

Mensink, W., Birrer, F. and Dutilleul, B. (2010). Unpacking European living labs: Analysing innovation's social dimensions. *Central European Journal of Public Policy* 4:1, 60–85.

Muchie, M. (2004). Resisting the deficit model of development in Africa: Re-thinking through the making of an African national innovation system. *Social Epistemology* 18:4, 315–332.

Nakasone, E., Torero, M. and Minten, B. (2014). The power of information: The ICT revolution in agricultural development. *Annual Review of Resource Economics* 6:1, 533–550.

Ndabeni, L., Rogerson, C. and Booyens, I. (2016). Innovation and local economic development policy in the global South: New South African perspectives. *Local Economy* 31:1–2, 299–311.

Nfila, R. and Jain, P. (2011). Managing indigenous knowledge systems in Botswana using information and communication technologies. In *Proceedings of 1st International Annual Conference of the Faculty of Communication and Information Science.* National University of Science & Technology (NUST), Islamabad, Pakistan, August 23–24.

Papaioannou, T. (2014). How inclusive can innovation and development be in the twenty-first century? *Innovation and Development* 4:2, 187–202.

Papaioannou, T. (2018). *Inclusive Innovation for Development: Meeting the Demands of Justice Through Public Action.* Routledge, London.

Paunov, C. (2013). Innovation and inclusive development: A discussion of the main policy issues. *OECD Working Papers* 2013/1. OECD, Paris.

Percy, C., Isaac, M. and Pamela, C. (2010). Understanding the relationship between indigenous (traditional) knowledge systems (IKS), and access to genetic resources and benefits sharing (ABS). *African Journal of Biotechnology* 9, 9204–9207.

Piketty, T. (2014). *Capital in the Twenty-First Century*. Harvard University Press, Cambridge, MA.

Planes-Satorra, S. and Paunov, C. (2017). *Inclusive Innovation Policies*. OECD, Paris.

Rakodi, C. (2016). The urban challenge in Africa. In Keiner, M., Koli-Schretzenmayr, M. and Schmid, W. (eds) *Managing Urban Futures: Sustainability and Urban Growth in Developing Countries*, 63–86. Routledge, London.

Republic of Rwanda (2019). *e-Soko*. www.esoko.gov.rw/. Retrieved June 2019.

Scheuermaier, U., Katz, E. and Heiland, S. (2004). *Finding New Things and Ways That Work. A Manual for Introducing Participatory Innovation Development (PID)*. LBL Swiss Center for Agricultural Extension, Lindau.

Schillo, R. and Robinson, R. (2017). Inclusive innovation in developed countries: The who, what, why and how. *Technology Innovation Management Review* 7:7, 34–46.

Shearmur, R., Carrincazeaux, C. and Doloreux, D. (eds) (2016). *Handbook on the Geographies of Innovation*. Edward Elgar, Cheltenham.

Sillitoe, P. and Marzano, M. (2009). Future of indigenous knowledge research in development. *Futures* 41, 13–23.

Sindzingre, A. (2013). The ambivalent impact of commodities: Structural change or status quo in Sub-Saharan Africa? *South African Journal of International Affairs* 20:1, 23–55.

Subba, R. (2006). Indigenous knowledge organization: An Indian scenario. *International Journal of Information Management* 26, 224–233.

Sun, Y. (2018). China's 2018 financial commitment to Africa: Adjustment and liberation. *Brookings*, September 5.

Taylor, I. (2015). Dependency redux: Why Africa is not rising. *Review of African Political Economy* 43, 8–25.

Trojer, L., Rydhagen, B. and Kjellqvist, T. (2014). Inclusive innovation processes – Experiences from Uganda and Tanzania. *African Journal of Science, Technology, Innovation and Development* 6:5, 425–438.

Twiga (2019). *Twiga*. www.twiga.ke/. Retrieved June 2019.

von Hippel, E. (2015). Democratizing innovation: The evolving phenomenon of user innovation. *Journal für Betriebwirtschaft* 55:1, 63–78.

von Hippel, E. and Katz, R. (2002). Shifting innovation to users via toolkits. *Management Science* 48:7, 821–833.

Weichselgartner, J. and Kasperson, R. (2010). Barriers in the science-policy-practice interface: Toward a knowledge-action-system in global environmental change research. *Global Environmental Change* 20, 266–277.

World Bank (2010). *Innovation Policy. A Guide for Developing Countries*. World Bank, Washington, DC.

World Bank (2019). *Taking the Pulse of Africa's Economy*. April 8. World Bank, Washington, DC.

World Health Organization (2006). Protecting traditional knowledge: The San and Hoodia. *Bulletin of the World Health Organization* 84.

Wyndberg, R., Schroeder, D. and Chennells, R. (eds) (2009). *Indigenous Peoples, Consent and Benefit Sharing: Lessons from San-Hoodia Case*. Springer, Berlin.

9

CONCLUSIONS

Introduction

Innovations for development are crucial for Africa and its future sustainable transformation as a continent from which poverty has been eradicated, where development is sustainable and where people can hope and see a promising future for them, their children and their grandchildren. In the 21st century, it is clear that Africa cannot rely on technology transfer from more developed countries and follow their science and technology policies and trajectories. Instead, Africa needs to look inside, turn its challenges into opportunities, unleash the creative potential of its population and create African innovation systems and policies. Such innovations need to be inclusive and create a framework for sustainable economic growth and development on the continent. Development cooperation focussed on inclusive innovations and participation can support African ambitions and practices for such a promising future.

In this concluding chapter, after this introduction, we focus on the key findings and suggestions of this book, *Innovation for Development in Africa*. In Section 9.2, we discuss what kinds of innovations and innovation processes are particularly relevant for Africa. For innovations, knowledge and its interpretation, application and transformation are needed. Locally specific indigenous knowledge can be an important asset to increase competitiveness and necessary smart specialisation of African localities. We also turn to innovation processes and highlight the opportunities deriving from different modes of knowledge and innovation creation in Africa. So far, in the establishment of innovation systems in Africa, not enough attention has been paid to the participatory processes.

In Section 9.3, we highlight the opportunities arising from development cooperation, taking innovation as its key element. Designed and implemented appropriately, innovation-focussed development cooperation is an instrument supporting suitable development paths and trajectories of African countries and communities. However, such aid and assistance play only a minor role in development. A major role in development needs to be filled by the African people and institutions themselves. In Section 9.4, we provide suggestions for future research and policymaking. Despite a commonly shared understanding that innovation is always context-specific, the majority of research and strategies do not consider enough differences between developing countries. They often link innovation development to science and technology, and innovations are seen to emerge linearly from science, research and development. From our viewpoint, innovation for development needs to be a continuing process and receive support and participation from below. For this, all stakeholders should be in interaction for sustainable and inclusive social and economic transformation of Africa.

- For sustainable development and beneficial social and economic growth in Africa, innovation development and innovation processes need to focus on issues that are particularly relevant for Africa.
- Development in Africa can be supported with development cooperation focussing on innovations, but it needs to involve both broader development goals and more concrete local actions that have been created reciprocally with aid and assistance donors and receivers.
- For inclusive innovations and transformative innovation systems, open-ended and agile innovation policies and research are needed to promote deep-learning processes and encourage experimentation, diversification and anticipation of the future.

Innovations and innovation processes in Africa

The emergence of innovations and the related innovation processes are in some aspects universal, but in other aspects, they are very context-specific. General commonalities are in the curiosity and motivation of individuals to search for new discoveries. Inventions turn into innovations if the institutional settings are suitable to help balance the supply and demand of innovation and various needed actors to interact together. The demand for innovations – enhanced products, services and processes – is obviously different in different development contexts, for example, in the countries in the Global North and those in Africa.

For the African contexts, three opportunities for innovations and innovation processes need to be utilised much more than they are today. The first opportunity is to consider inclusiveness in all innovation development and innovation processes. In such a way, innovations would be more targeted to specific African contexts and needs. The possibility of remembering inclusiveness can be enhanced by broadening the participation into innovation processes and keeping innovation processes open. New platforms, such as inclusive living labs, are needed to enhance such processes. As we have demonstrated in this book, almost all successful innovations emerging from Africa are inclusive and built to solve existing social challenges. Inclusive innovation and innovation systems are also about modes and ethics of innovation (Bryden et al., 2017). Inclusiveness also means discovering niche markets for innovations and potentially wider beneficiaries when innovations are scaled up among innovation developers and innovation users. However, inclusive innovation does not always have to mean economic growth (Schillo & Robinson, 2017). "Designed and made in Africa" could be a trademark that guarantees that inclusiveness is taken into account and that Africa, its people and its enterprises would benefit. Inclusiveness could then become an important aspect of corporate responsibility from small, early-stage start-ups to established large corporations regarding both African enterprises and other enterprises operating on the continent.

The second opportunity is about processes leading to innovations. As mentioned in this book, many innovations in the economically more advanced and societally more developed Global North rely on improvements in science and technology, solid institutions with supportive long-term innovation policies, substantial R&D investments and advanced knowledge creation in globally leading universities. Integrating such STI-based innovation into national or local strategies is not demanding but turning them into well-functioning practices. The Global South and Africa in particular cannot base their innovation development and innovation systems on similar processes like the more developed areas of the world do. Africa lacks knowledge, actors, finance and policies for such innovations. In addition, these STI-based innovations often consider inclusiveness, inequality and poverty external to innovation processes, although those aspects need to be at the heart of the African innovation system.

African innovation systems could more carefully consider their main strength in innovation development, namely the growing population, generally much younger population structure, rapid urbanisation and the burning need for new challenge-related and user-centric innovations. With suitable combinations, the African innovation scene could become prolific and fast-developing, always searching for new incremental and radical methods as well as end products and services to meet the growing demand. In addition, being able to succeed in local African contexts could mean a possibility to scale up the products and services to other contexts in Africa, in the Global South and beyond. There are many possibilities for rising African innovators and innovations that can make a global breakthrough. In an increasingly competitive world, a competitive advantage in Africa could arise from applying indigenous knowledge in the innovation processes and its end products and services. Incorporating aspects of indigenous knowledge into user-centric innovations would mean that such products and services could not be copied and easily produced elsewhere with lower costs. Furthermore, indigenous knowledge is more than a source of new product innovations. It is also a way to engage local communities in the innovation development, as many times, indigenous knowledge is the most well-known knowledge in these communities (Weichselgartner & Pigeon, 2015).

The synthesis of innovation processes in Africa discovered in this book are presented as the innovation matrix in Figure 9.1. We developed it to better contextualise and recognise the key actors and knowledge of innovation

	Private sector	**HUB**	**HEI**	**NGO/DC**	
STI Tech-enabled	Existing products to emerging middle-class consumers	High-tech start-ups	Development of science-based hard and soft technology	Frugal and responsible innovation	Market push How to localise?
DUI Need-driven	Innovations from local needs	Community-based innovations	Innovations developed in community-university cooperation	Inclusive innovation process	Market pull How to scale?
	Efficiency, growth, corporate governance	Connecting, dynamic, concrete, flexible	Project- and program-driven, multiple stakeholders, rigid	Unique element, transformative, sustainability?	

FIGURE 9.1 Innovation matrix in Africa

processes in Africa. It distinguishes innovations along the STI mode and the DUI mode of learning and knowledge generation. In the matrix are four central actors in African innovation development: the private sector, co-creation hubs, higher education institutes and non-governmental organisations/donor communities. The STI and DUI modes are combined by different central actors, resulting in eight types of innovation processes. These in turn result in different types of innovations that each require specific political and operational measures. The matrix can be applied as an open analytical tool to analyse local innovation policies or as an open policy tool to design more accurate innovation policies. This division inside the innovation processes is not exclusive, and several innovations develop in the interactions between various actors.

The third aspect is to recognise that African countries are behind many countries in the world as regards their general development stage. The reduction of this gap is important, but the trajectories to do it cannot be based on following and imitating the more developed Global North and hoping that the technology transfer will substantially reduce the development gap. There are no magic shortcuts in development, but African countries do not need to follow all the steps the more developed countries took in the past 50 years when innovations emerged and became constituted as the key development factors there.

The pathway of Africa can also be diverse based on local generation, adaptation and experimentation of knowledge. Knowledge is the fundament of innovations, but not always analytical knowledge from science. More applied synthetic knowledge takes advantage of the tacit indigenous knowledge and combines it with applicable knowledge that, for example, many African universities can provide. Universities can become integral parts of inclusive innovation systems in Africa (see Arocena et al., 2017). In addition, learning from long-term sustainable practices of indigenous cultures, traditions and people might create an advantage in inserting the necessary symbolic knowledge in the design and marketing of the final products and services in a such way that these turn into commercial successes in Africa and beyond.

Development cooperation focussing on innovations in Africa

Africa has received and continues to receive substantial development assistance from outside the continent. The OECD-DAC countries continue to be the main development donors, but new South-South development cooperation is rapidly rupturing the customary nexus. The target countries are usually those that are the least developed. In addition, there are donors outside the Global North governments, such as large international institutions. The World Bank, the IMF and various UN agencies (UNECA, UNDP, UNESCO and UNICEF, to name a few) are significant institutions of global reach that support the development of Africa. Furthermore, there are many global foundations annually supporting various aspects of development in the continent with tens of millions of USD.

Commonly, donors also have interests in their aid provisions (i.e. what kind of Africa should emerge as the result of their interventions). Such global reach activities have become even more important because donors outside the Global North have become active in Africa. China is one, and perhaps the clearest, example to illustrate how donors are not present in Africa just to help but also to satisfy their own business interests.

Similarly, the once top-down international development aid to Africa has transformed into development cooperation. Cooperation should mean that the donor and the receiver mutually agree on the cooperation scopes, targets and impacts. However, in the current development cooperation, such mutual interests are not always so evident. Many donor countries have turned their development policies into more selfish practices with the aim to support their domestic enterprises to gain markets in developing countries. Private sector involvement has brought new actors, knowledge and money to the cooperation, but it has mutually blurred the ultimate objectives and target groups of such cooperation. Oftentimes, the beneficiaries of the private sector-led development are not the poorest communities, but the well-off minorities located especially in the capitals of the Global South.

Furthermore, the contemporary development cooperation is based firmly on faith that what the market decides is, in the end, beneficial for everyone. The aim is to support the partner in the Global South, including Africa, to become more entrepreneurial and unleash the creativity and capacity building into economic success and growth of individuals, enterprises and countries through innovations. To put it simply, the global poor should lift themselves from poverty. Development aid and cooperation provide these people with related knowledge, skills and entrepreneurial mentality, and commercial innovations are one crucial tool for development. Therefore, innovations and support for the emergence of innovations also need to be evaluated as a social and political process.

The development cooperation between the Global North and Africa should be a true cooperation in which partners are equal. The development goals should be decided through broad participation, and the practices should also engage those whom this aid is intended to help. This would increase the ownership of development for those to whom the projects' impact is targeted, and supposedly, it would also make the largest impact. However, development (aid) policies are always political and ideological as well. Therefore, particular attention should be paid to the nature and goals of cooperation. Instead of seeing entrepreneurship and markets to decide the viability of development, the goals should be to raise equality, reduce poverty and guarantee long-term sustainability of development with varied trajectories and modes of action. Therefore, an impact assessment needs to be carried out before any development actions and placing substantial weight on globally and locally significant goals. The vested interests of the donors should be considered and exposed, and ways should be found to create win-win situations in which the benefits are not primarily, or at least always, counted in terms of economic turnover and profits. The goals should also consider development issues toward the latter part of the 21st century and beyond.

Suggestions for further research and policymaking

This book, *Innovation for Development in Africa*, encourages broader interest toward innovations, innovation-focussed development policies and Africa. Our intention has been to give an introduction to such a wide topic. Development policies focussing on change through innovations are a rather new phenomenon in the relations between the Global North and Africa, so the impact of such policies is still to be seen. While we urge creation of evidence-based development policies, we did not want to wait another decade to provide research-based evidence about the impacts. Instead, we have shed light on international development policies and cooperation and how these meet and interact with innovations in Africa.

Our point of view is that action needs to be taken now in research, policies and practices. Not just any action is needed, but those actions that help in the eradication of poverty, provision of equality, democratisation of societies and communities with broad and deep participatory processes and better fulfilment of the various needs of Africans – these are key elements in supporting the achievement of long-term sustainability on the continent. The time is now – compared with today, in a few decades, there will be more than a half billion more people living in African cities. Many people need to leave home, and some also the continent, in search of a better life and livelihood. As we have underlined, the world's future depends on Africa. If development in Africa is positive, the world might see a prosperous 22nd century. If development fails in Africa, the risks are large that no one will see the 22nd century or that the world will be completely different in terms of justice, equality and fulfilment. Our viewpoint is positive: Africa can undergo a positive social and economic transformation that leads to the long-term sustainability of the continent and meaningful lives of its numerous inhabitants.

So far, there has been only limited research on the intertwining of innovation, innovation policies and systems, and international development policies, especially regarding Africa. Therefore, all studies are more than welcome. Many actors are involved in innovation-focussed development – the public, private, and non-governmental sectors, universities and other development actors, just to name a few. All these topics require research projects that can cover the whole continent or selected African countries or in-depth analyses that focus on individual projects. Innovation and innovation system development are crucial for Africa, so singular, multiple, longitudinal, future-oriented, comparative, case-based, etc. research about them is needed. In particular, the grand challenges need to be addressed and equally wise solutions need to be found and implemented. The research needs to show possibilities and pitfalls for inclusive innovation and inclusive innovation systems in Africa and their impact on African societies, local communities and individuals. One important aspect is to strengthen the research capacity of African researchers in innovation and knowledge creation studies. For this, research cooperation is needed, regardless of the country or continent of the researchers.

One distinctive factor of African innovation systems is the active role of the international development community as a catalyst and supporter of this development. Innovations' role in general, and in particular in international development aid and cooperation, has not yet been sufficiently addressed. There is a need for both rapid research interventions and long-term longitudinal analyses because international development policies transform rapidly. New instruments, actors and tools emerge in development cooperation, resulting in processes leading to inclusion and exclusion of the African population, as well as resulting in less and more sustainable development.

What about the policy suggestions for innovation for development in Africa? Well, as the title of this book indicates, innovations are needed for development in Africa. We suggest that policies in Africa be open-ended, experimental and diverse; we see this book also as an open policy tool that acknowledges local variety concerning the main actors and the knowledge-based of innovation development. We are conscious of our limitations in addressing all regional knowledge in this book. As we mentioned previously, we do not think that policy transfer is possible – it is difficult to achieve success by imitating and forcefully placing institutions and instruments from the Global North in various African countries and contexts. Of course, one can learn about others' projects and processes.

So, is there an African way for development? The answer is that there are various African ways for development with innovations. Setting goals and implementing the policies, programs and projects is also about ideologies and politics. Global sustainability cannot be omitted, nor the short-term and long-term progress of Africa and its population. The African population continues to grow, so more added value needs to be created in Africa, and such added value needs to be divided more equally among its inhabitants. It is timely, and even urgent, for policymakers' frame, but also, experimenting with new policy practices can be tried to solve these challenges. No one should be doomed to live in poverty and constrained by structural injustice that still impedes many Africans' ability to develop and to rise from misery.

Although the DUI mode of innovation might be more significant for inclusive and sustainable societal development, African policies do not need to totally reject the STI mode. Instead, it is essential to develop agile policies that recognise diverse actors and modes of knowledge and, most essentially, the various combinations of those. Open-ended policies promote deep-learning processes and encourage experimentation, diversification and anticipation of the future. In addition to related variety, African countries could utilise open-ended policies to create unrelated variety in the global knowledge society (Frenken, 2017). Instead of aiming to catch up and imitating the policy transformation of the Global North, creating innovation systems based on their comparative advantages and contributing to transformative change can lead to transformative processes and mutual deep-learning processes between the Global South and Global North that will be beneficial for all (Diercks et al., 2019).

We believe in African people, individuals and communities; in their incredibly rich creativity, hopes and traditions; and in their capacities to absorb innovations and create novel knowledge and innovations. We wish to see the prosperous future of the continent. Inclusive innovations keep Africa rising.

- Successful intertwining of innovations, innovation processes and development in Africa is crucial for the continent and the whole world.
- Research-based results support evidence-based policies regarding innovation-focussed development cooperation, so there is a need for rapid research interventions and long-term longitudinal analyses on the backgrounds, practices and impacts of innovation-focussed development cooperation.
- Inclusive innovations and transformative innovation policies are key to reaching a sustainable Africa.

Discussion questions

- What is a transformative innovation system in Africa?
- What opportunities and challenges exist to reach a sustainable Africa with the help of transformative innovation systems?
- Discuss how research about development and innovations in Africa can support the achievement of sustainable development in Africa.

References

Arocena, R., Göransson, B. and Sutz, J. (2017). *Developmental Universities in Inclusive Innovation Systems. Alternatives for Knowledge Democratization in the Global South.* Springer, Berlin.

Bryden, J., Gezelius, S., Refsgaard, K. and Sutz, J. (2017). Inclusive innovation in the bioeconomy: Concepts and directions for research. *Innovation and Development* 7:1, 1–16.

Diercks, G., Larsen, H. and Steward, F. (2019). Transformative innovation policy: Addressing variety in an emerging policy paradigm. *Research Policy* 48:4, 880–894.

Frenken, K. (2017). A complexity-theoretic perspective on innovation policy. *Complexity, Innovation and Policy* 3:1, 35–47.

Hooli, L., Jauhiainen, J., Järvi, A., Nkonoki, E., Taajamaa, V. and Käyhkö, N. (2019). Contextualising innovation in Africa: Knowledge modes and actors in local innovation development. In *IST-Africa Week Conference (IST-Africa) Proceedings*, Nairobi, Kenya, 2019.

Schillo, R. and Robinson, R. (2017). Inclusive innovation in developed countries: The who, what, why and how. *Technology Innovation Management Review* 7:7, 34–46.

Weichselgartner, J. and Pigeon, P. (2015). The role of knowledge in disaster risk reduction. *International Journal of Disaster Risk Science* 6:2, 107–116.

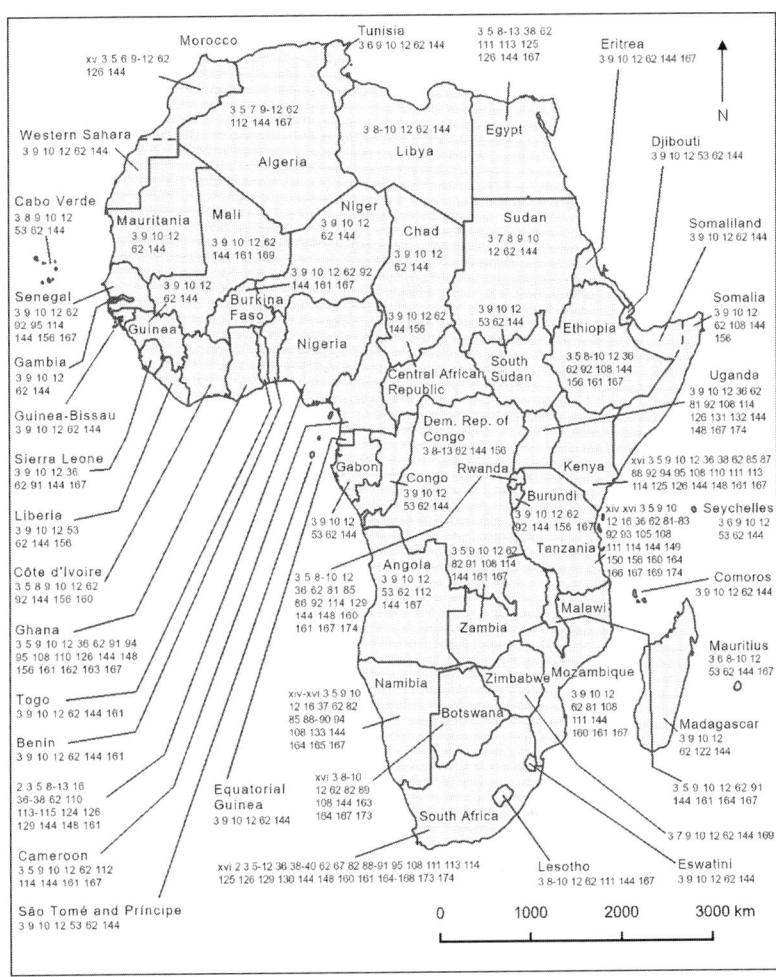

Morocco
xv 3 5 6 9-12 62
126 144

Tunisia
3 6 9 10 12 62 144

3 5 8-13 38 62
111 113 125
126 144 167

Eritrea
3 9 10 12 62 144 167

3 5 7 9-12 62
112 144 167

3 8-10 12 62 144
Libya

Egypt

Djibouti
3 9 10 12 53 62 144

Western Sahara
3 9 10 12 62 144

Algeria

Cabo Verde
3 8 9 10 12
53 62 144

Mauritania
3 9 10 12
62 144

Mali
3 9 10 12 62
144 161 169

Niger
3 9 10 12
62 144

Chad
3 9 10 12
62 144

Sudan
3 7 8 9 10
12 62 144

Somaliland
3 9 10 12 62 144

Senegal
3 9 10 12 62
92 95 114
144 156 167

3 9 10 12
62 144

Burkina
Faso

3 9 10 12 62 92
144 161 167

Nigeria

3 9 10 12 62
144 156

3 9 10 12
53 62 144

Ethiopia

Somalia
3 9 10 12
62 108 144
156

Guinea

Gambia
3 9 10 12
62 144

Central African
Republic

South
Sudan

3 5 8-10 12 36
62 92 108 144
156 161 167

Uganda
3 9 10 12 36 62
81 92 108 114
126 131 132 144
148 167 174

Guinea-Bissau
3 9 10 12 62 144

Sierra Leone
3 9 10 12 36
62 91 144 167

Dem. Rep. of
Congo
3 8-13 62 144 156

Kenya

xvi 3 5 9 10 12 36 38 62 85 87
88 92 94 95 108 110 111 113
114 125 126 144 148 161 167

Liberia
3 9 10 12 53
62 144 156

Gabon

Congo
3 9 10 12
53 62 144

Rwanda

Burundi
3 9 10 12 62
92 144 156 167

xiv xvi 3 5 9 10
12 16 36 62 81-83
92 93 105 108
111 114 144 149
150 158 160 164
166 167 169 174

Seychelles
3 6 9 10 12
53 62 144

Côte d'Ivoire
3 5 8 9 10 12 62
92 144 156 160

3 9 10 12
53 62 144

3 5 9 10 12 62
82 91 108 114
144 161 167

Tanzania

Comoros
3 9 10 12 62 144

Ghana
3 5 9 10 12 36 62 91 94
95 108 110 126 144 148
156 161 162 163 167

Angola
3 9 10 12
36 62 81 85
86 92 114 129
144 148 160
161 167 174

3 9 10 12
53 62 112
144 167

Malawi

Mauritius
3 6 8-10 12
53 62 144 167

Togo
3 9 10 12 62 144 161

Zambia

Mozambique
3 9 10 12
62 81 108
111 144
160 161 167

Benin
3 9 10 12 62 144 161

Namibia

Zimbabwe

Botswana

Madagascar
3 9 10 12
62 122 144

2 3 5 8-13 18
36-38 62 110
113-115 124 126
129 144 148 161

xiv-xvi 3 5 9 10
12 16 37 62 82
85 88-90 94
108 133 144
164 165 167

3 5 9 10 12 62 91
144 181 164 167

Cameroon
3 5 9 10 12 62 112
114 144 161 167

Equatorial
Guinea
3 9 10 12 62 144

xvi 3 8-10
12 62 82 89
108 144 163
164 167 173

South Africa

3 7 9 10 12 62 144 169

Eswatini
3 9 10 12 62 144

São Tomé and Príncipe
3 9 10 12 53 62 144

xvi 2 3 5-12 36 38-40 62 67 82 88-91 95 108 111 113 114
125 126 129 130 144 148 160 161 164-168 173 174

Lesotho
3 8-10 12 62 111 144 167

0   1000   2000   3000 km

N

INDEX

9780367349561